Climate Engineering

Climate Engineering

A Normative Perspective

Daniel Edward Callies

LEXINGTON BOOKS
Lanham • Boulder • New York • London

Published by Lexington Books
An imprint of The Rowman & Littlefield Publishing Group, Inc.
4501 Forbes Boulevard, Suite 200, Lanham, Maryland 20706
www.rowman.com

6 Tinworth Street, London SE11 5AL, United Kingdom

Copyright © 2019 by The Rowman & Littlefield Publishing Group, Inc.

All rights reserved. No part of this book may be reproduced in any form or by any electronic or mechanical means, including information storage and retrieval systems, without written permission from the publisher, except by a reviewer who may quote passages in a review.

British Library Cataloguing in Publication Information Available

Library of Congress Cataloging-in-Publication Data

Names: Callies, Daniel Edward, 1986- author.
Title: Climate engineering : a normative perspective / Daniel Edward Callies.
Description: Lanham : Lexington Books, [2019] | Includes bibliographical references and index.
Identifiers: LCCN 2019019767 (print) | LCCN 2019020592 (ebook) | ISBN 9781498586689 (electronic) | ISBN 9781498586672 (cloth) | ISBN 9781498586696 (pbk.) Subjects: LCSH: Climate change mitigation--Moral and ethical aspects. | Climate change mitigation--Government policy. | Stratospheric aerosols. | Environmental justice.
Classification: LCC TD171.75 (ebook) | LCC TD171.75 .C35 2019 (print) | DDC 363.738/746--dc23
LC record available at https://lccn.loc.gov/2019019767

For my father, who always is there for me, and in memory of my mother, who always was there for me.

Contents

Acknowledgments — ix

1. Introduction — 1
2. Research — 23
3. Deployment — 51
4. Legitimacy — 75
5. Substantive Justice — 93
6. Procedural Justice — 115
7. Conclusion — 137

Bibliography — 151
Index — 165
About the Author — 169

Acknowledgments

My largest intellectual debt is owed to the time I spent at Johann Wolfgang Goethe University in Frankfurt, Germany. Frankfurt boasts a truly exceptional academic environment, and I consider myself lucky to have been a part of it. Specifically, I am indebted to Professor Darrel Moellendorf. Darrel held the chair of international political theory during my stay in Frankfurt, and he's had an immeasurable influence on my development as a philosopher. But I am also deeply indebted to him for his timely, thorough, and challenging comments on every chapter of this book.

In addition to the beer gardens filled with pilsners and pretzels, some of my fondest memories in Frankfurt are of the remarkable political theory colloquium at Goethe. Having attended various colloquia throughout the world, I can truly say that Frankfurt's stands out. I want to thank Professor Rainer Forst both for creating the environment in which that colloquium has flourished, and for providing me with helpful comments on drafts of this book. Professor Forst has had a huge impact on my thinking, something that comes out in Chapter 6.

In addition to the colloquium, I also had countless stimulating discussions and interactions with the many PhD students, postdocs, visitors, and faculty at Goethe University. Special thanks to: Ayelet Banai, Mahmoud Bassiouni, Thomas Biebricher, Idil Boran, Julian Culp, Cristian Dimitriu, Dimitris Efthimiou, Corrado Fumagalli, Dorothea Gädeke, Andrew Greetis, Malte Ibsen, Beth Kahn, Bruno Leipold, Merielli Mafra, Esther Lea Neuhann, Antoinette Scherz, Regina Scheidel, Johannes Schulz, Rosa Sierra, Anita Silveira, Abraham Singer, and Luke Ulaş.

I'd also like to thank everyone from the amazing International Political Theory Lehrstul at Goethe (AKA: Team Moellendorf): Nate Adams, Susanne Boerner, Maria Paola Ferretti, Daniel Hammer, Brian Milstein, and

Tatjana Visak, and especially Merten Reglitz, Eszter Kollar, and Lukas Sparenborg. I've profited from numerous lunches, coffees, and late-night conversations with you all, and I am deeply grateful for it. I should also mention the generous funding and research leaves provided to me by Frankfurt's Excellence Cluster: The Formation of Normative Orders, which was funded by the German Research Foundation (Deutsche Forschungsgemeinschaft).

Outside of Frankfurt, I've also benefited from productive stays at Harvard's Kennedy School of Government and the Université catholique de Louvain's Hoover Chair. I thank Axel Gosseries and Danielle Zwarthoed for the invitation to stay in Louvain-La-Neuve. From Harvard, I want to thank: Eric Beerbohm, Lizzie Burns, Tweedy Flanigan, Joshua Horton, Peter Irvine, David Keith, Jeroen Oomen, and Gernot Wagner.

I've workshopped chapters of this book at a number of academic conferences, the participants of which I owe many thanks. Specifically, thanks to those who provided food for thought at: the International Society for Environmental Ethics, Anchorage (2018); The Morality and Policy of Negative Emissions, Senckenberg Natural History Museum (2018); Geoengineering and Legitimacy Conference, University of Washington (2017); Secondary Climate Injustices Workshop, University of Warwick (2017); Warwick Graduate Conference on Political and Legal Theory, University of Warwick (2016); Harvard Political Theory Workshop, Harvard University (2016); Science, Technology, and Public Policy Workshop, Harvard University (2016); Climate Scholars' Conference, University of Reading (2016); ISOE: Institute for Social-Ecological Research, Frankfurt (2016); Climate Ethics and Economics Workshop, Goethe University Frankfurt (2016); and the International Society for Environmental Ethics, Kiel (2015).

While we may have met at conferences, I'd like to thank academics with whom I've had significant contact over the past five years and who have shaped my thinking on matters of environmental ethics in general. Thanks to: Christian Baatz, Marion Hourdequin, Dominic Lenzi, Catriona McKinnon, Lukas Meyer, Kian Mintz-Woo, David Morrow, Konrad Ott, Christopher J. Preston, Jesse Reynolds, Dominic Roser, Henry Shue, Harald Stelzer, Allen Thompson, and Ivo Wallimann-Helmer. And special thanks go to Clare Heyward, Toby Svoboda, Benjamin Hale, Jamie Draper, Stephen Gardiner, and Augustin Fragniere for either in-depth discussions or comments on drafts of chapters.

A version of Chapter 4 of this book appears in *Ethics, Policy, and Environment*, and my discussion of the slippery slope argument in Chapter 2 was published in the *Journal of Applied Philosophy*. I am grateful both to the editors and publishers of these journals for permission to use the work that appeared there. I'd like to thank to Jana Hodges-Kluck and Trevor Cowell at Lexington for guiding me through the publication process, and I owe thanks to the anonymous reviewer they found who pushed me to revise various parts of the manuscript.

Finally, I am forever indebted to my friends and family for putting up both with me and my absence while working on this project. Specifically, thanks to the Moellendorf/Friedmann family for taking me in upon arriving in Frankfurt, and thanks to Nick, Joachim, and especially Ellen Nieß for making me a part of their family (and extra thanks to Ellen for solving just about every problem I ever encountered in Germany). Most importantly, thanks to my dad and brothers who have helped me and supported me from afar. Their sacrifices made writing this possible.

While the aforementioned people have undoubtedly had a positive impact on this project, it goes without saying that any errors or shortsightedness are solely mine. I'm sure there are countless others whom I should be naming here. Please know that you have my gratitude.

Chapter One

Introduction

§1 A WARMING WORLD

The science has been rather clear for decades—the world is warming, rapidly, and we're the cause. Our everyday activities, seemingly innocuous in isolation, are combining to have profound effects on the planet. Despite having known about these profound effects for decades, we as a global community have been rather lethargic in responding to the problem of climate change through the two most appropriate measures available: mitigation and adaptation. With this lethargy around mitigation and adaptation in mind, the previously-considered taboo subject of geoengineering has emerged as a prospective policy response to climate change.

Defined as the "deliberate, large-scale manipulation of the planetary environment in order to counteract anthropogenic climate change,"[1] geoengineering—or climate engineering—is an umbrella term encapsulating a wide array of technological proposals. One of these proposals in particular—one that goes by the name of stratospheric aerosol injection (SAI)—has received quite a bit of attention from natural scientists and engineers who tend to recognize its potential to allay some of the harms associated with climate change.[2] However, many normative theorists have tended to view the technology suspiciously, highlighting unethical aspects of research and deployment[3] and the technology's inherently troubling political implications.[4] The purpose of this book is to evaluate certain ethical and political aspects of intentionally manipulating the climate by means of SAI.

While SAI is not currently being used to offset climate change, there are active research programs around the world investigating the technology. Does engaging in such research perhaps lead us down a slippery slope toward inexorable deployment? Could it be that even researching such a technology will draw us away from the more important tasks of mitigation and adaptation? Should we err on the side of caution and avoid risky interventions in the climate system altogether? Surely, if we were to engage in research and potential deployment of a climate engineering technology, governance would be a must. But what would count as legitimate governance? Moreover, research and deployment of the technology will most likely create a novel distribution of benefits and burdens. What should we consider a just distribution of these benefits and burdens? And, perhaps most importantly, who ought to be included in the decision-making process surrounding geoengineering and what should that process look like? These are some of the questions this book attempts to address.

§2 THE TRAGEDY OF CLIMATE CHANGE

Now, one might be tempted to ask: *How did we even get to the point of considering the intentional manipulation of the planetary environment?* Since at least 1990, we've known about the threat of global climate change. The Intergovernmental Panel on Climate Change's First Assessment Report told us (with certainty) that emissions resulting from human activities are substantially increasing atmospheric concentrations of greenhouse gases, and that "these increases will enhance the greenhouse effect, resulting on average in an additional warming of the Earth's surface."[5] This warming of the Earth's surface—which has already hit nearly 1°C (compared to pre-industrial averages) and could reach up to 4.8°C by the end of the century[6] —carries with it significant negative corollaries. For example, we have already begun and will continue to lose certain species that are simply unable to cope with the rate at which average temperatures are increasing, both on land and in the oceans.[7] This loss of biodiversity will threaten the health of many of our natural ecosystems. But the threats of climate change are not confined to natural ecosystems—human systems are at risk as well. From agriculture to infrastructure, climate change will have significant negative aggregate effects, with those negative effects hitting underdeveloped nations and the least well-off members of all nations the hardest.[8]

Recognition of the threats posed by climate change notwithstanding, emissions have continued to rise nearly every year for the past century. And this is not because people doubt the imminent threat posed by a rapidly changing climate. Given our current stage of technological development, burning fossil fuels such as coal, oil, and natural gas is one of the cheapest ways of producing energy. And energy is necessary for bringing people out of crippling poverty and for maintaining the comfortable lifestyles of many around the globe. This is one of the features that makes climate change such a pernicious problem. While everyone may recognize the need for global action on climate change, no one wants to give up their access to cheap energy.

This explanation of climate change has led many to consider the problem a kind of "tragedy of the commons."[9] The idea of the tragedy of the commons was first introduced by William Foster Lloyd,[10] but was made popular by Garrett Hardin in his essay by the same name.[11] A tragedy of the commons occurs when a shared resource is destroyed or depleted through a process whereby the individual actors sharing the resource rationally pursue their own self-interest to the detriment of the group as a whole. In the classic formulation, a common pasture is shared by a group of herdsman. It is in the self-interest of each herdsman to add a head of cattle to his herd since he receives the entire benefit of doing so, while the cost of the extra cattle is shared by all. The benefit is conceived as the price the herdsman will get for the head of cattle at the market and the cost is the additional negative impact on the shared pasture. Eventually, with each herdsman acting rationally on his own self-interest, the pasture's carrying capacity is surpassed and the land is no longer able to support any of the cattle, bringing about devastation for all.

The parallel to climate change and greenhouse gas emissions should be clear. We all share the atmosphere in common and rely upon it for a stable climate. We can think of the herdsman as individuals or small collectives of current people. It is in the interest of each individual or small collective to increase their emissions and reap the associated benefits (while exporting the costs to humanity—and the planet—as a whole). That is, absent some collective agreement, it is rational for each individual or collective to continue emitting. The tragedy, which has already begun, is the overloading of the atmosphere with heat-trapping gasses, bringing with it unprecedented warming and potentially catastrophic climatic effects.

Fortunately for us—and even more fortunate for future generations—we have a collective agreement to address this structural problem. In December 2015, 195 signatories representing every person on the globe[12] committed to collectively limiting warming to no more than 2°C.[13] In November 2016, after the required minimum number of signatories had ratified the text, the agreement went into effect. Contrary to its predecessor the Kyoto Protocol—which was designed top-down with each state receiving its emission allotment after determining an appropriate upper limit—the Paris Agreement was an example of bottom-up international cooperation. Each party, with the overall goal of limiting warming to 2°C in mind, individually and independently determined what it was willing to contribute—or, more accurately, how much it was willing to limit future emissions. While the Paris Agreement is a monumental achievement of international collaboration and represents a significant first step toward avoiding dangerous interference with the planetary environment, it is widely recognized to be too little and too late.

Under the agreement, parties pledge initial mitigation goals and then revise these goals every five years. Unfortunately, the pledges are going to have to get significantly more ambitious if they are to have any hope of limiting climate change to 2°C. For instance, if all parties were to fully satisfy their current Paris commitments and do nothing more, the world would experience a warming of roughly 3°C.[14] And this, of course, is assuming that all parties actually meet their commitments—a rather dubious assumption, especially after President Trump's announcement of his intention to withdraw the United States from participation in the agreement.[15] It is this bleak prognosis regarding mitigation and adaptation that has resparked the academic discussion on geoengineering.

§3 GEOENGINEERING: CDR AND SRM

Of course, the idea of manipulating weather is not a new one. The ancient Greeks made sacrifices to Zeus, the Romans pleaded with Jupiter for precipitation, and the Mojave tribe of the Southwestern United States adorned headdresses and turquois during ceremonial dances performed to induce rain for their crops. Given our long-standing interest in controlling the weather, present-day proposals of climate engineering aimed at alleviating the effects of global warming should come as no surprise.

What may come as a surprise to some is that no one is *currently* engineering the climate. According to an international survey, roughly one in six people believe it to be true or partly true that chemicals are currently being sprayed from passenger jets as part of a sinister plot by governments.[16] These dangerous "chemtrails," the conspiracy theorists claim, are released with the goal of population control or even the psychological manipulation of the masses. Still others think that such chemtrails—which are actually just condensation trails, or "contrails," left by regular passenger jets flying through the right atmospheric conditions—are being used for weather or climate modification. This conspiracy theory is just that: a conspiracy theory roundly rejected by atmospheric scientists and aviation experts.[17] It is unfortunate that many people think of chemtrails when they think of the term "geoengineering," but the focus of this book lies elsewhere.

If it isn't chemtrails that we are talking about when we say "geoengineering," what is it? Definitions of geoengineering vary slightly. But, in general, when people speak of geoengineering they have something like the previously-quoted Royal Society definition in mind. Geoengineering refers to a number of different specific technologies that all aim at "deliberate, large-scale manipulation of the planetary environment in order to counteract anthropogenic climate change."[18] These technologies are often grouped into two categories: (1) carbon dioxide removal (CDR) projects, which aim to remove

CO_2 from the atmosphere; and (2) solar radiation management (SRM) projects, which aim to reflect a small percentage of incoming solar radiation back out into space.

Carbon dioxide removal projects can range from the seemingly mundane to the grandiose. Afforestation, or the large-scale, intentional planting of trees, lies somewhere toward the mundane end of the spectrum. Planting trees is a project familiar to just about everyone, and this familiarity conditions our initial response to its use as a solution to climate change. However, there are risks that accompany even afforestation. The major risks arise primarily due to the immense amount of land it would take to meaningfully affect climate change through tree-planting alone. We need our arable land for agriculture, one, and, two, covering the earth's surface with forests could negatively impact biodiversity.[19] Not only that, but given the time it takes trees to sequester carbon dioxide, afforestation may not be able to help in the short-term.[20] Still, afforestation is gradual and is considered to be less risky (though, of course, less efficient) than other forms of CDR.

A proposal that lies somewhere closer to the grandiose end of the spectrum would be massive fertilization of the oceans with iron. There is a natural cycle within our oceans in which photosynthesizers at the surface transfer carbon to the deep sea. Algae, for instance, takes in substantial amounts of CO_2 and, when its remnants sink to the deep sea, it is consumed by bacteria and other organisms. These organisms convert the algal remnants back into CO_2, thus completing the transfer of atmospheric CO_2 into "deep ocean CO_2." By introducing large amounts of iron into the open ocean, we could create large algal blooms, enhancing this natural biological pump that "buries" atmospheric CO_2 into the deep ocean. Of course, there are significant risks associated with ocean fertilization. Due to the significant uncertainty surrounding the workings of various marine ecosystems, the exact side-effects of large-scale ocean fertilization are unknown. For billions of people around the world—especially the least well-off members of the global community—the oceans are a lifeline, which makes experimenting with them dangerous.[21]

The associated risks notwithstanding, some CDR projects will almost certainly have to be part of the "toolkit" we use to respond to climate change. In fact, almost every IPCC scenario that limits warming to below 2°C requires some negative emissions technology.[22] And while some prefer CDR methods to SRM since the former are seen as addressing the *cause* of climate change as opposed to merely its *effects*, there are still significant ethical and political issues associated with most forms of CDR. Carbon dioxide removal, however, is not the focus of this book. We will be primarily concerned with solar radiation management.

Like CDR, Solar Radiation Management proposals range from the mundane to the grandiose. Toward the mundane end of the spectrum lies the proposal to enhance the reflectivity of human environments. By, for instance, painting the roofs of our buildings with lighter colors and changing our roadways from a heat-trapping black to a heat-reflecting white, we can enhance the earth's albedo effect. With more radiation reflected back into space, the earth is a little cooler than it otherwise would be. The main problem with relying upon the enhancement of the albedo of human settlements for responding to climate change is that it would be (a) expensive, since we would have to continually re-paint the chosen surfaces, (b) slow-acting, since it would take us a while to actually convert our roads and rooftops, and (c) dramatically insufficient, since most of the earth is not covered with roads and buildings.

If enhancing the reflectivity of human settlements is insufficient, perhaps we need to think bigger. What is probably the most grandiose of SRM techniques is the proposal to place mirrors in low-earth orbit so as to deflect solar radiation before it even reaches the lower atmosphere. While such a proposal has the potential to significantly reduce the amount of radiation reaching Earth's surface (and, thus, significantly cool the planet), the costs of launching the needed number of mirrors into orbit is essentially prohibitive.[23]

§3.1 Stratospheric Aerosol Injection

Somewhere in between the mundane and the grandiose—though, perhaps closer to the grandiose—lies the technology that is the focus of this book, stratospheric aerosol injection. The main idea behind the proposal, as its name implies, is to inject aerosols into the stratosphere. These aerosols would create a semi-permeable layer capable of shielding the planet from some of the incoming solar radiation. Of course, the less radiation that makes it to Earth's surface, the less radiation there is to be trapped by the greenhouse effect.

There are a number of different delivery systems to release the aerosols in the stratosphere that are currently being discussed. For instance, we could use military-grade artillery guns to project the aerosols or weather balloons outfitted with long hoses that would reach back down to the earth's surface. Perhaps the most popular and one of the most cost-effective of delivery methods involves the use of regular business jets. A Boeing 747 or fleet of similar aircraft could continually deploy 1 metric ton (Mt) of aerosols at the required altitude, enough to offset roughly half of the expected temperature increase due to anthropogenic global warming.[24]

Along with the different delivery systems, there are also various kinds of aerosols that could be used. The most viable option at the moment is some kind of sulfate aerosol, either sulfur dioxide (SO_2) or hydrogen sulfide (H_2S). One of the primary advantages to using sulfate aerosols is our understanding of their effect. In 1991, Mount Pinatubo released somewhere between 10–20 million tons of sulfur into the atmosphere that resulted in an average global cooling of 0.5°C for the year.[25] The injection of sulfate aerosols into the stratosphere would mimic this natural volcanic effect.

Perhaps the two greatest merits of SAI are its rapid efficacy and its comparative cost.[26] First, once introduced into the stratosphere, the aerosols would start producing the desired cooling effect within weeks. In comparison, emissions mitigation will only have a cooling effect across a time span of decades or centuries due to the inertia of the climate system. This near immediate efficacy is a significant benefit of the proposal. Second, the annual cost of releasing the aforementioned 1 Mt of aerosols into the stratosphere with retrofitted business jets could be as little as $1 billion.[27] One billion USD per year may sound like a lot. But, when compared to either the cost of damages from unchecked climate change or the cost of emissions mitigation needed to avoid climate change, $1 billion per year is as close to free as it comes.[28] However, it should be stressed that while SAI is comparatively cheap, it is not by any means a perfect substitute for mitigation. This is for at least the following four reasons.[29]

First, while a certain temperature might be equally achievable through emissions mitigation as through geoengineering, temperature and climate are different things. Achieving the same temperature target through geoengineering as would come about by just emitting fewer greenhouse gases will result in novel climate configurations, most notably affecting perception patterns. Second, reaching a certain temperature target through geoengineering rather than via emissions reductions may have significant effects on air pollution and atmospheric ozone concentrations. Third, achieving a certain temperature target through emissions mitigation will go a long way toward solving the problem of ocean acidification. Reaching the same temperature via geoengineering will lead to a world with dramatically more acidic oceans. And, fourth, there are unique risks ("unknown-unknowns") associated with geoengineering that are not associated with emissions mitigation. Reducing our emissions will leave us with an atmosphere and a climate similar to the one of our recent past, whereas implementing a geoengineering scheme will create novel, perhaps harmful, climatic and atmospheric conditions for some. So, while it is true that the price tag associated with geoengineering research and deployment is comparatively modest, it is a mistake to see geoengineering as a perfect substitute for emissions mitigation.

Before moving on to some thoughts on methodology, I want to offer a quick note on terminology. As we have just seen, there are significant differences between the various proposals that all fall under the heading of "geoengineering." This being the case, it is often important to specify the particular geoengineering proposal one is discussing when making empirical and normative statements about it.[30] For stylistic purposes, I'll be loose with language throughout this project. Thus, whenever reference is made to geoengineering, solar geoengineering, climate engineering, SRM, or SAI, it is the specific proposal of stratospheric aerosol injection that I have in mind (unless otherwise specified).

§4 NORMATIVE THEORIZING

Now, with this natural-scientific explanation of geoengineering in mind, you might be wondering what a philosopher would have to say about such a technical issue. Isn't geoengineering something that should be researched by natural scientists and engineers? What could so-called armchair theorizing contribute to the debate? The fact of the matter is that, despite seeming rather complicated from a technical standpoint, the thorniest problems related to climate engineering are normative, not empirical. An empirical or descriptive analysis of the ethics of geoengineering might, for example, poll people or engage public deliberative groups to uncover what people *actually* think about the ethics of intentionally manipulating the planet. An empirical or descriptive analysis of the politics or governance of geoengineering might be conducted by scholars in the field of international relations, economics, or public policy who could offer descriptive claims about how the technologies might *actually* be governed. In contrast to these kinds of analyses, the analysis of geoengineering conducted in this book will be normative—that is to say, we will be looking at what people *should* think about the ethics of geoengineering and we'll explore concepts and principles that *ought* to guide our thinking about geoengineering governance. So, we can say that the purpose of this project is to offer guidance to our normative thinking about climate engineering.

§4.1 Reflective Equilibrium

With this purpose of the project in mind, the natural question is: How can one arrive at sound moral and political judgments about controversial subjects such as climate engineering? Even further, how can one arrive at sound moral and political judgments in general? There are various different methods we can employ when making normative judgments. We may simply rely upon our intuitions in deciding right from wrong—an approach known aptly as *intuitionism*. Or we might derive the answers to moral questions from religious texts, like the Bible or the Quran. These approaches, however, have serious difficulties in dealing with moral disagreement. People's intuitions about moral issues are notoriously at odds with one another, and the same goes for those who adhere to different religious traditions.

Perhaps the best normative methodological approach, and the one I adopt here, is the method championed by John Rawls—that of *wide reflective equilibrium*.[31] Wide reflective equilibrium can be contrasted with narrow reflective equilibrium. Narrow reflective equilibrium can be described as a situation in which one has reached an equilibrium point between one's considered moral judgments and a set of moral principles that account for those judgments.[32] When employing the *method* of narrow reflective equilibrium, we start by identifying certain moral judgments, and then select those about which we are most confident (e.g., the judgment that slavery is wrong).[33] Those moral judgments about which we are most confident—our *considered* moral judgments—are those made under favorable conditions, that is, the person judging has adequate information, is calm, etc. We then look for a set of principles that can best account for our considered moral judgments. Sometimes there will be discrepancies between principles and considered moral judgments. Thus, the final step of the method of narrow reflective equilibrium is to attempt to achieve just that, an equilibrium between the particular judgments and the set of principles that account for them. Ideally, we would go back and forth, rejecting particular moral judgments when the principles they are at odds with enjoy greater support, or revising our principles when they clash with (strong) enough considered moral judgments. This point of coherence between our considered moral judgments and our set of moral principles is narrow reflective equilibrium.

This *narrow* reflective equilibrium, however, is not enough—what we want is *wide* reflective equilibrium. The problem with narrow reflective equilibrium is twofold. First, even under favorable conditions, our considered moral judgments may be the product of historical accident, bias, or ideology.[34] Finding a proper fit between (unjustified) considered judgments and moral principles would not amount to much of an epistemological justification. Second, there may be various different sets of principles that fit our considered moral judgments. Wide reflective equilibrium addresses these shortcomings. Wide reflective equilibrium is a method for arriving at an acceptable coherence between (a) our considered moral judgments, (b) a set of moral principles, and (c) relevant background theories. This addition of a third category allows us to dispel certain biased moral judgments and allows us to justifiably choose between sets of principles that similarly "fit" our moral judgments. Norman Daniels writes,

> we advance philosophical arguments intended to bring out the relative strengths and weaknesses of the alternative sets of principles. . . . These arguments can be construed as inferences from some set of relevant background theories. . . . Assume that some particular set of arguments wins and that the moral agent is persuaded that some set of principles is more acceptable than the others. . . . We can imagine the agent working back and forth, making adjustments to his considered judgments, his moral principles, and his background theories. In this way he arrives at an equilibrium point that consists of the ordered triple (a), (b), (c).[35]

Along these lines, wide reflective equilibrium offers a somewhat independent criterion capable of both (1) jettisoning biased moral judgments, and (2) selecting between competing sets of principles that account for our considered moral judgments.[36] When making moral arguments throughout this project, it is this method upon which I am relying.

§4.2 Ideal and Nonideal Theory

Even if we are in agreement about how to do moral theory, we may be in disagreement about how to do political theory. There are different methodological approaches one might take. For example, there is significant discussion among political philosophers and theorists as to whether one should engage in more abstract ideal theorizing about justice and legitimacy, or whether we should be more realistic and give weight to the constraints of our nonideal world. The discussion in the literature regarding ideal and non-ideal theorizing, like the discussion surrounding the method of reflective equilibrium, owes its genesis to Rawls. In laying out his conception of liberal egalitarian justice, Rawls confines his analysis to the domain of what he dubs "ideal theory," which sets an ideal or a target of justice for society to aim at.[37] Under the primary assumption of strict compliance—meaning that all relevant agents fully comply with the principles of justice—the main purpose of ideal theorizing is to paint the picture of a perfectly just social arrangement. The assumption that everyone will comply with the recognized principles of justice is, of course, no small assumption. But, by working out what ideal justice demands, we can then analyze where and how much current arrangements are falling short of this ideal.[38] Rawls contrasts ideal theory with "nonideal theory," which, as opposed to assuming full compliance, assumes only partial compliance (and, thus, partial noncompliance) with the demands of full justice.[39]

Today, there are at least three different understandings of the ideal/nonideal distinction. The first is the one attributed to Rawls—the distinction between full compliance and partial compliance. The second is the distinction between "utopian" or "idealistic" theory, on the one hand, and "realistic" theory, on the other. The third understanding sees the distinction as one of "end-state" theory and "transitional" theory.[40]

In later chapters of this book, I rely upon both ideal and nonideal methodology. For instance, Chapter 4 puts forward a conception of institutional legitimacy in which an institution can be justifiably considered legitimate if it satisfies enough normative criteria to a sufficient degree. The fact that an institution does not have to fully satisfy or strictly comply with any of the relevant normative criteria in order to be considered legitimate means that we can understand the conception put forward as an example of nonideal theorizing of the Rawlsian variety.[41] Though it could also be understood in the two other ways nonideal theory is conceived. We could consider this conception of legitimacy as both more "realistic" since it doesn't require the onerous complete fulfillment of the identified normative criteria to render the institution legitimate. And it could be said that, in this sense, it is a "transitional" theorization of the concept of legitimacy since it recognizes that legitimate (though not fully just) institutions may be necessary to make progress on justice.[42] With this conception of legitimacy in mind, Chapters 5 and 6 then put forward ideal conceptions of two of those normative criteria that a geoengineering governance institution ought to approximate if it is to be considered legitimate. Specifically, for an institution overseeing geoengineering to be considered legitimate, it ought to conform to norms of procedural and distributive justice. Chapters 5 and 6 outline what it would take for these criteria to be fully realized, without requiring their full realization for a positive legitimacy assessment. Thus, both methodologies of ideal and nonideal theory are relied upon.

§5 STATE OF RESEARCH

It was not until recently that the realistic possibility of intentionally manipulating the climate actually materialized. Thus, compared to many other topics, academic research focusing on geoengineering is still in a fledgling stage. The contemporary scientific discussion around geoengineering could perhaps be dated to 1974 with the publication of Kellog and Schneider's "Climate Stabilization: For Better or for Worse?"[43] However, as is commonly referenced, there was somewhat of a taboo around the topic in the final decades of the twentieth century. It was not until the 2006 publication of Nobel Prize–winning chemist Paul Crutzen's article that geoengineering really became a serious part of the discussion regarding climate policy.[44] Since then, there has been an abundance of literature focusing on the science and technical feasibility of prospective ways in which we can engineer the climate.

Following along with the technical literature, there have been a number of academics within law and economics who have broached the subject. For example, Daniel Bodansky, Albert Lin, John Virgoe, Edward Parson, and David Victor have all discussed geoengineering governance from the perspective of international law.[45] Scott Barret's influential piece on the "incredible" economics of geoengineering spurred others within the discipline to engage with the topic.[46] Juan Moreno-Cruz has analyzed the strategic interaction of countries on the international stage once geoengineering becomes a viable policy option.[47] And Elizabeth Burns, Gernot Wagner, and others have looked at public opinion with respect to the very idea of engineering the climate.[48]

There have even been a few commissioned reports by various governments, nonprofits, and scientific academies.[49] The most comprehensive report on the subject to date would probably have to be the Royal Society's 2009 *Geoengineering the Climate: Science, Governance, and Uncertainty*.[50] Comprising the expertise of physicists, engineers, lawyers, environmental scientists, and policy experts, the report described several geoengineering proposals and evaluated them according to standards of effectiveness, timeliness, affordability, and safety. Of course, while not surprising given the report's authorship, the two-page discussion on the ethics of geoengineering left something to be desired. Aiming to fill this ethical gap, the Oxford Geoengineering Programme published its influential "Oxford Principles" paper in *Climatic Change*.[51] According to the authors, geoengineering research ought to be guided by five norms: (1) Geoengineering should be regulated as a global public good; (2) There should be public involvement in decision-making; (3) There should be disclosure of the results of research; (4) There should be an independent assessment of the potential impacts of any research; and (5) There should be robust governance prior to any deployment. Despite their far-reaching impact, the social scientists at Oxford were not the first to look at geoengineering from an ethical perspective.

The first philosopher to address the ethical implications of geoengineering was Dale Jamieson in 1996.[52] Despite presenting numerous ethical considerations that seemed to condemn geoengineering, Jamieson ultimately concluded that research should go forward as long as such research were to take into account the ethical concerns he presented. Following Jamieson, philosophers have addressed different ethical aspects related to intentionally manipulating the climate. Perhaps the most prolific is Stephen Gardiner, who generally makes ethical arguments concerning the context in which geoengineering is discussed. For instance, Gardiner argues that we should avoid calling geoengineering either a "public good" or a "plan b," insisting instead that it is not a public good (in the canonical sense) and that it represents more of a "plan z."[53] And, similarly, he argues against the rationale that we should research geoengineering now in the event that it is ever needed in the future.[54] He suggests such reasoning may simply be an instance of what he calls "moral corruption."[55] We'll look at some of Gardiner's arguments in the next chapter.

Many others have advanced philosophical inquiry on geoengineering as well. For instance, remaining within the realm of the right context for debate, David Morrow argues (contra Gardiner) that not only does SAI meet the canonical definition of a global public good, but that it is also useful to look at the technology within such a framework.[56] Looking at the technology as a global public good, Morrow posits, allows us to draw upon useful literature in other fields such as economics and political science. Konrad Ott has brought the discussion on geoengineering to the intergenerational realm. Ott argues that if one generation were to deploy SRM, it might be placing future generations in a moral dilemma. It is not unrealistic, Ott argues, to envision SRM carrying with it harms of its own, even if it delivers net benefits. If this turns out to be the case, then we could be placing future generations in a moral dilemma in which they either (a) have to continue using SRM and thus continue producing the accompanying harm it entails, or (b) have to make the decision to halt SRM deployment, which could produce disastrous side effects as well.[57]

Of course, our ethical evaluation of geoengineering could very well depend upon our particular conception of morality. Different theories grant moral standing to different groups of agents. Whether we are anthropocentrists, biocentrists, or ecocentrists, for instance, may very well influence our ethical evaluation of a particular technology. In "The Ethics of Geoengineering," Toby Svoboda analyzes SRM from anthropocentric, animal liberationist, and biocentric ethical outlooks, concluding that the technology poses prima facie problems for all three.[58]

Perhaps some of the most thoughtful work on the ethics of geoengineering has come from Clare Heyward and Christopher J. Preston. In her sensible and sober book chapter titled "Is There Anything New Under the Sun?: Exceptionalism, Novelty, and Debating Geoengineering Governance,"[59] Heyward aims to provide reasonable parameters for a constructive debate about climate engineering. She advocates not for a particular position with respect to technological responses to climate change, but rather contends that the debate ought to be conducted without reference to claims of exceptionalism, that is, claims that climate engineering "will lead to unprecedented situations, either good or bad."[60] Heyward highlights that climate engineering proponents often rely on the "climate emergency" justification, a justification that appeals to an exceptionally bad climate catastrophe in which we might be willing to engineer the planet as the lesser of the two evils. Likewise, opponents of climate engineering often claim that such technologies are ungovernable or at least incompatible with democratic governance, a claim Heyward shows as exceptional and doubtful. She concludes that both sides of the climate engineering debate would do well to avoid such exceptional claims.

In addition to having edited two anthologies on climate engineering,[61] Christopher J. Preston is responsible for having written an influential early article that laid out the ethical landscape for the geoengineering discussion. In his "Ethics and Geoengineering: Reviewing the Moral Issues Raised by Solar Radiation Management and Carbon Dioxide Removal,"[62] Preston highlights a number of ethical issues related to geoengineering according to the "temporal space" in which they would occur. For instance, some ethical concerns crop up with respect to research, some with respect to deployment, and still others crop up when consideration cessation. Preston's schematic provides a rough framework that influenced my thinking on this project.

The words of the previous paragraphs notwithstanding, work on the ethics and politics of geoengineering has barely begun to scratch the normative surface. In comparison to the abundance of scientific literature, philosophy and political theory are lagging far behind; and this on a topic for which normative theorizing is of the utmost importance. There is currently only one monograph in moral and political theory that focuses on the ethics, legitimacy, and justice of geoengineering specifically.[63] This book aims to add a fresh view to the developing ethical literature on geoengineering, and fill the gap that exists in political theory.

§6 PROJECT OUTLINE

The project is divided into seven chapters. Chapters 2 and 3 are devoted to ethical concerns, analyzing various arguments pertaining to research and deployment of the technology. In Chapter 2 I look at four arguments related to researching SAI: the arm the future argument and the economic argument *in favor* of research; and the slippery slope argument and the moral hazard argument *against* research. The arm the future argument claims that we should research geoengineering now in case it is ever needed in the future. I look at Stephen Gardiner's analysis of the argument and offer both critique and restrained support. The economic argument claims that we ought to research geoengineering because of how incredibly cheap it is compared to climate change damages and mitigation. Despite highlighting the shortcomings of geoengineering as a substitute for mitigation, I claim that the cheap price tag does, in fact, offer a reason to go forward with research. The slippery slope argument says (1) that if we engage in research, it will inexorably lead to deployment, (2) we don't want to deploy the technology, (3) therefore, we shouldn't engage in research. I argue that premise one (the empirical premise) and premise two (the normative premise) are both doubtful, rendering the argument unsound. The moral hazard argument says that we shouldn't engage in research because it will weaken our resolve to mitigate and adapt to climate change. I offer a number of reasons to doubt this argument. The two most prominent reasons would be that the link between geoengineering research and a weakened resolve for mitigation is empirically lacking, and that even if our resolve to mitigate did weaken, we should want to know whether that loss in resolve to mitigate is not offset by the potential benefits that such research would yield. The ultimate conclusion of Chapter 2 is that neither of the arguments surveyed provide us with sufficient reason to abandon research.

Next, in Chapter 3, I look at the precautionary argument, the respect for nature argument, and the playing God argument against the deployment of the technology. The precautionary argument against deployment builds out of the precautionary principle, ultimately claiming that we should err on the side of caution and avoid the risks associated with geoengineering, even if we are uncertain as to what those effects are and what the probabilities of them occurring amount to. Drawing upon Sunstein's analysis of the precautionary principle, I argue that the prudent or cautious decision must be chosen from the risk/risk scenario we currently find ourselves in (the risk of deploying geoengineering versus the risk of allowing unchecked climate change to continue). Under this framing, it is unclear which way the precautionary principle cuts with respect to geoengineering research and deployment. I then examine another decision-rule for situations with uncertain outcomes—the minimax rule. While minimax reasoning would ground a moratorium on geoengineering, I argue that the conditions that make minimax a reasonable decision-rule don't hold when it comes to geoengineering and climate change.

The respect for nature argument claims that even if deployment of geoengineering would prove beneficial for mankind and the planet as a whole, such intentional intervention in the climate system necessarily shows a contempt or disrespect for nature. Through an exploration of what exactly it means to show an attitude of "respect" for nature, I argue that geoengineering may in fact amount to us dominating nature, which is one way of showing disrespect. However, I argue that it is doubtful that such domination is of a different kind than we have been engaged in for centuries, and that it is difficult for us to imagine a world in which we don't dominate nature. Thus, the deployment of geoengineering may not be as much of an affront to nature as is imagined, or it may not exemplify an impermissible kind of disrespect.

The playing God argument relies upon the doctrine of doing and allowing, saying that since deployment will save some from climate change but may condemn others to a premature death due to unavoidable side effects of the technology, deployment is impermissible since it amounts to deciding who lives and who dies. I cast doubt upon the absolutist form of the doctrine of doing and allowing at the state (and international) level, and propose that the doctrine of double effect could serve as a possible justification for the decision to deploy. The conclusion of Chapter 3 is that these arguments fail to show that deployment of geoengineering would necessarily be impermissible. Thus, the ultimate conclusion of the first three chapters of the book is that the ethical arguments against geoengineering cannot serve to ground a moratorium and that research should continue. That conclusion notwithstanding, it is recognized that there is a need for oversight and regulation of research, development, and any potential deployment in order to assure that the negative prospects outlined in Chapters 2 and 3 don't materialize.

The subsequent chapters of the book take the need for oversight to heart. The next three chapters focus on political aspects of geoengineering, making the case in favor of an institution to oversee research and deployment that is governed by norms of legitimacy, distributive justice, and procedural justice. Chapter 4 begins with an analysis of the concept of legitimacy. I argue that conceptions of state legitimacy are ill-suited to assess a geoengineering regulatory institution due to significant dissimilarities between the function and power of the two kinds of institutions. I develop a broad account of institutional legitimacy drawing upon Allen Buchanan's recent Metacoordination View, and propose five normative criteria to serve, not as necessary and sufficient conditions, but as considerations that would speak in favor of a geoengineering institution's legitimacy.

Having outlined a conception of legitimacy, Chapters 5 and 6 provide some content to two specific normative criteria: substantive justice and procedural justice. Chapter 5 looks at the normative criterion of substantive justice that ought to guide geoengineering governance. I put forward a convergence thesis. I argue that three common sense considerations of fairness all point to the same conclusion about how the benefits and burdens of geoengineering should be distributed. The fact that the least well-off members of the global community bear the least causal responsibility for climate change, bear the least beneficiary responsibility for climate change, and have the weakest ability to respond to the threats of climate change all point to the conclusion that they have the strongest claims to the benefits (and against the burdens) associated with geoengineering. In other words, if geoengineering is to be researched, developed, and deployed, substantive justice demands that it should be used primarily to benefit the least well-off members of the global community.

But that substantive outcome is not all we care about; we also care about *how* the decisions surrounding geoengineering are made. Chapter 6 thus focuses on the normative criterion of procedural justice. Just processes often have both instrumental and intrinsic value. They have instrumental value in that they can secure desirable outcomes, and intrinsic value in that they allow us to both (a) fulfil our natural duties of justice and (b) relate to one another in a morally valuable way. Having outlined the reasons we are concerned with procedural justice, the remainder of the chapter is devoted to finding a suitable general principle for procedural justice as it relates to geoengineering. After surveying the all-affected principle and the Equal Influence Principle, I argue that the Proportionality Principle does the best job of justifiably providing fair terms of participation for an SRM decision-making process.[64] The chapter concludes by working out the possible implications the Proportionality Principle has for the real-world decision-making process around SRM, having outlined a conception of legitimacy and provided some content to the normative criteria of substantive and procedural justice. Chapter 7 offers some concluding remarks, with a short discussion on nonideal situations, intergenerational justice, and anthropocentric ethics.

§7 FINAL REMARKS

For good reason, geoengineering is a controversial subject that seems to divide scholars. I am often asked whether I am "for" or "against" the idea of engineering the climate. I genuinely find it unfortunate that geoengineering is being seriously considered as a policy response to climate change. This is not because I think there is something reprehensible or hubristic about intentionally altering the climate—though, that may be the case. The reason I find it unfortunate that we are currently considering the merits of geoengineering as a response to climate change is that it indicates that we have failed in our primary climate obligations. One thing that almost everyone in the geoengineering debate agrees about is that mitigation and adaptation are the two best tools in our policy toolkit. Many philosophers agree that we have a collective duty to respond to climate change. Had we taken our duty seriously by aggressively pursuing mitigation and more heavily investing in adaptation, we very well may have found ourselves in a world in which geoengineering wasn't even on the table. That, regrettably, is not the world in which we find ourselves.

The world in which we find ourselves in one beset with ethical predicaments. Given our failing with respect to mitigation and adaptation, it is important that we get the ethics and politics of geoengineering right. This book is an attempt at doing just that. It is doubtful that you will agree with every one of my conclusions. While I certainly do hope to convince you, I would consider this project a success if, at the very least, it advances the debate. And it is to that debate that we now turn.

NOTES

1. Royal Society, *Geoengineering the Climate: Science, Governance and Uncertainty* (London: Royal Society, 2009), 1.

2. Paul J. Crutzen, "Albedo Enhancement by Stratospheric Sulfur Injections: A Contribution to Resolve a Policy Dilemma?" *Climatic Change* 77, no. 3-4 (September 1, 2006): 211–20, https://doi.org/10.1007/s10584-006-9101-y; David W. Keith and Douglas G. MacMartin, "A Temporary, Moderate and Responsive Scenario for Solar Geoengineering," *Nature Climate Change* 5, no. 3 (February 16, 2015): 201–6, https://doi.org/10.1038/nclimate2493; Ralph J. Cicerone, "Geoengineering: Encouraging Research and Overseeing Implementation," *Climatic Change* 77, no. 3-4 (September 1, 2006): 221–26, https://doi.org/10.1007/s10584-006-9102-x.

3. See Stephen Gardiner, "Is Arming the Future with Geoengineering Really the Lesser Evil?," in *Climate Ethics*, ed. Stephen Gardiner et al. (Oxford: Oxford University Press, 2010), 284–312.

4. See Bronislaw Szerszynski et al., "Why Solar Radiation Management Geoengineering and Democracy Won't Mix," *Environment and Planning A* 45, no. 12 (December 2013): 2809–16, https://doi.org/10.1068/a45649.

5. Intergovernmental Panel on Climate Change, *Climate Change: The IPCC Scientific Assessment*, ed. John Theodore Houghton, G. J. Jenkins, and J. J. Ephraums (Cambridge: Cambridge University Press, 1990), xi.

6. This is the high end of RCP8.5, commonly considered the "business as usual" scenario in which we do not reduce our emissions. See Intergovernmental Panel on Climate Change, *IPCC, 2014: Summary for Policymakers. In: Climate Change 2014: Impacts, Adaptation, and Vulnerability. Part A: Global and Sectoral Aspects. Contribution of Working Group II to the Fifth Assessment Report of the Intergovernmental Panel on Climate Change* (Cambridge: Cambridge University Press, 2014), 23, http://www.ipcc.ch/pdf/assessment-report/ar5/wg2/ar5_wgII_spm_en.pdf.

7. See Alan Pounds, Michael Fogden, and John Campbell, "Biological Response to Climate Change on a Tropical Mountain," *Nature* 398 (April 15, 1999): 611–15, as cited in Chris D. Thomas et al., "Extinction Risk from Climate Change," *Nature* 427, no. 6970 (January 8, 2004): 145–48.

8. Intergovernmental Panel on Climate Change, *Climate Change 2001: Synthesis Report*, ed. R. T. Watson et al. (Cambridge: Cambridge University Press, 2001).

9. For a critique of the idea that climate change is accurately characterized as a tragedy of the commons, see Stephen Gardiner, *A Perfect Moral Storm: The Ethical Tragedy of Climate Change* (New York: Oxford University Press, 2011).

10. William Foster Lloyd, "W. F. Lloyd on the Checks to Population," *Population and Development Review* 6, no. 3 (September 1980): 473–96, https://doi.org/10.2307/1972412.

11. Garrett Hardin, "The Tragedy of the Commons," *Science* 162 (December 1968): 1243–48.

12. The only two countries that were not signatories to the Paris Agreement were Syria, which was and still is in the middle of a prolonged civil war, and Nicaragua, which felt the agreement was not doing nearly enough to address climate change.

13. In fact, there was language included in the Paris Agreement that expressed a commitment to limit warming to 1.5°C.

14. Joeri Rogelj et al., "Paris Agreement Climate Proposals Need a Boost to Keep Warming Well below 2°C," *Nature* 534, no. 7609 (June 29, 2016): 631–39, https://doi.org/10.1038/nature18307.

15. For a critique of President Trump's announcement, see Daniel Edward Callies, "Paris Agreement Bigger than Any One Man," *Agenda for International Development*, June 2017.

16. A. M. Mercer, D. W. Keith, and J. D. Sharp, "Public Understanding of Solar Radiation Management," *Environmental Research Letters* 6, no. 4 (October 1, 2011): 044006, https://doi.org/10.1088/1748-9326/6/4/044006.

17. Christine Shearer et al., "Quantifying Expert Consensus against the Existence of a Secret, Large-Scale Atmospheric Spraying Program," *Environmental Research Letters* 11, no. 8 (2016): 084011, https://doi.org/10.1088/1748-9326/11/8/084011.

18. Royal Society, *Geoengineering the Climate: Science, Governance and Uncertainty*, 1.

19. Royal Society, 10.

20. Of course, carbon sequestered in trees is also vulnerable to being released by droughts, fires, and intentional deforestation.

21. Food and Agriculture Organization of the United Nations, *The State of the World Fisheries and Aquaculture 2014: Opportunities and Challenges* (Rome: Food and Agriculture Organization of the United Nations, 2014).

22. T. Gasser et al., "Negative Emissions Physically Needed to Keep Global Warming below 2°C," *Nature Communications* 6 (August 3, 2015): 7958, https://doi.org/10.1038/ncomms8958.

23. Royal Society, 33.

24. Keith and MacMartin, "A Temporary, Moderate and Responsive Scenario for Solar Geoengineering."

25. Mike Hulme, *Can Science Fix Climate Change? A Case against Climate Engineering*, New Human Frontiers Series (Cambridge: Polity Press, 2014), 44–45.

26. We'll look more at the cost of SAI in Chapter 2.

27. Keith and MacMartin, "A Temporary, Moderate and Responsive Scenario for Solar Geoengineering."

28. The estimates of both the costs of damages from climate change and the costs of mitigation vary widely. However, a justifiable estimate from Sir Nicholas Stern puts the costs of unchecked climate change at 5 percent of GDP annually and the costs of mitigation at 1 percent of GDP annually. William Nordhaus, espousing a more conservative estimate, assumes that unchecked climate change will cost us about 2.5 percent of global GDP by the end of the century. See Nicholas Stern, *The Economics of Climate Change: The Stern Review*, ed. Great Britain (Cambridge: Cambridge University Press, 2007); William D. Nordhaus, *A Question of Balance: Weighing the Options on Global Warming Policies*, 2015.

29. There are other reasons that SAI is not a perfect substitute for mitigation, but these three are perhaps the most prominent.

30. See Clare Heyward, "Situating and Abandoning Geoengineering: A Typology of Five Responses to Dangerous Climate Change," *PS: Political Science & Politics* 46, no. 1 (January 2013): 23–27, https://doi.org/10.1017/S1049096512001436.

31. Reflective equilibrium is not confined to the moral realm. It could be used as a theoretical method of justification in logic or mathematics as well—and some even see an analog of reflective equilibrium in the natural sciences. There is, of course, controversy as to whether reflective equilibrium is the best approach to moral theorizing. See Norman Daniels, "Reflective Equilibrium," in *The Stanford Encyclopedia of Philosophy*, ed. Edward N. Zalta, Winter 2016 (Metaphysics Research Lab, Stanford University, 2016), https://plato.stanford.edu/archives/win2016/entries/reflective-equilibrium/.

32. John Rawls, *Justice as Fairness: A Restatement* (Cambridge, MA: Harvard University Press, 2001), 29.

33. Moral judgments may be about any level of generality. To quote Thomas Scanlon, considered judgments "may be judgments about the rightness or wrongness of particular actions, general moral principles, or judgments about the kind of considerations that are relevant to determining the rightness of actions." See Thomas Scanlon, *Being Realistic about Reasons* (Oxford: Oxford University Press, 2014).

34. Norman Daniels, "Wide Reflective Equilibrium and Theory Acceptance in Ethics," *The Journal of Philosophy* 76, no. 5 (1979): 258, https://doi.org/10.2307/2025881.

35. Daniels, 258–59.

36. There are, of course, objections to even wide reflective equilibrium. Some utilitarians argue that the method of reflective equilibrium is akin to a fancy kind of moral intuitionism. For instance, Peter Singer argues that reflective equilibrium amounts to a kind of subjectivism. See Peter Singer, "Sidwick and Reflective Equilibrium," *The Monist* 58, no. 3 (1974): 490–517, https://doi.org/10.2307/27902380. I think Rawls' move to wide reflective equilibrium—which he didn't make until after Singer's article came out—goes some way toward avoiding this objection, but I don't take that up here.

37. This is similar to how Rawls' ideal theory is understood by Allen Buchanan. See Allen Buchanan, *Justice, Legitimacy, and Self-Determination* (Oxford: Oxford University Press, 2003), 34.

38. John Rawls, *A Theory of Justice*, 8.

39. There are actually two assumptions embedded in ideal theorizing. One is the assumption of full compliance with the demands of justice. The second is the idea that historical and material conditions are favorable to a society being able to achieve justice. See John Rawls, *The Law of Peoples* (Cambridge, MA: Harvard University Press, 2002), 5.

40. Laura Valentini, "Ideal vs. Non-Ideal Theory: A Conceptual Map," *Philosophy Compass* 7, no. 9 (September 1, 2012): 654–64, https://doi.org/10.1111/j.1747-9991.2012.00500.x.

41. For example, contrast the conception of institutional legitimacy put forward with a voluntarist conception of legitimacy in which an institution is legitimate if and only if it enjoys the actual voluntary consent of all those over whom it wields power. This voluntarist conception of legitimacy would be an ideal conception.

42. See Chapter 4, Section 4.1.

43. W. W. Kellogg and S. H. Schneider, "Climate Stabilization: For Better or for Worse?" *Science* 186, no. 4170 (December 27, 1974): 1163–72, https://doi.org/10.1126/science.186.4170.1163.

44. Crutzen, "Albedo Enhancement by Stratospheric Sulfur Injections."

45. Daniel Bodansky, "The Who, What, and Wherefore of Geoengineering Governance," *Climatic Change* 121, no. 3 (December 2013): 539–51, https://doi.org/10.1007/s10584-013-0759-7; Albert C Lin, "Geoengineering Governance," *Issues in Legal Scholarship* 8, no. 1 (January 13, 2009), https://doi.org/10.2202/1539-8323.1112; D. G. Victor, "On the Regulation of Geoengineering," *Oxford Review of Economic Policy* 24, no. 2 (June 1, 2008): 322–36, https://doi.org/10.1093/oxrep/grn018; John Virgoe, "International Governance of a Possible Geoengineering Intervention to Combat Climate Change," *Climatic Change* 95, no. 1–2 (July 1, 2009): 103–19, https://doi.org/10.1007/s10584-008-9523-9; Edward A. Parson, "Climate Engineering in Global Climate Governance: Implications for Participation and Linkage," *Transnational Environmental Law* 3, no. 1 (April 2014): 89–110, https://doi.org/10.1017/S2047102513000496.

46. Scott Barrett, "The Incredible Economics of Geoengineering," *Environmental and Resource Economics* 39, no. 1 (January 1, 2008): 45–54, https://doi.org/10.1007/s10640-007-9174-8.

47. Juan Moreno-Cruz, "Mitigation and the Geoengineering Threat," *Resource and Energy Economics* 41, no. C (2015): 248–63.

48. Elizabeth T. Burns et al., "What Do People Think When They Think about Solar Geoengineering? A Review of Empirical Social Science Literature, and Prospects for Future Research," *Earth's Future* 4, no. 11 (November 1, 2016): 536–42, https://doi.org/10.1002/2016EF000461.

49. Lee Lane et al., "Workshop Report on Managing Solar Radiation," April 1, 2007, https://ntrs.nasa.gov/search.jsp?R=20070031204; Jane Long, "Task Force on Climate Remediation Research," October 4, 2011, https://bipartisanpolicy.org/library/task-force-climate-remediation-research/; U. S. Government Accountability Office, "Climate Change: Preliminary Observations on Geoengineering Science, Federal Efforts, and Governance Issues," no. GAO-10-546T (March 18, 2010), http://www.gao.gov/products/GAO-10-546T; "The European Transdisciplinary Assessment of Climate Engineering (EuTRACE)," accessed August 17, 2017, https://www.adelphi.de/en/publication/european-transdisciplinary-assessment-climate-engineering-eutrace.

50. The Royal Society, at p. 1.

51. Steve Rayner et al., "The Oxford Principles," *Climatic Change* 121, no. 3 (December 2013): 499–512, https://doi.org/10.1007/s10584-012-0675-2.

52. Dale Jamieson, "Ethics and Intentional Climate Change," *Climatic Change* 33, no. 3 (July 1, 1996): 323–36, https://doi.org/10.1007/BF00142580.

53. Stephen M. Gardiner, "Why Geoengineering Is Not a 'Global Public Good', and Why It Is Ethically Misleading to Frame It as One," *Climatic Change* 121, no. 3 (December 1, 2013): 513–25, https://doi.org/10.1007/s10584-013-0764-x.

54. Stephen M. Gardiner, "The Desperation Argument for Geoengineering," *Political Science & Politics* 46, no. 1 (January 2013): 28–33, https://doi.org/10.1017/S1049096512001424.

55. Gardiner, *A Perfect Moral Storm*.

56. David R. Morrow, "Why Geoengineering Is a Public Good, Even If It Is Bad," *Climatic Change* 123, no. 2 (March 2014): 95–100, https://doi.org/10.1007/s10584-013-0967-1.

57. Konrad Ott, "Might Solar Radiation Management Constitute a Dilemma?," in *Engineering the Climate: The Ethics of Solar Radiation Management*, ed. Christopher J. Preston (Lanham, MD: Lexington Books, 2012), 113–31.

58. Toby Svoboda, "The Ethics of Geoengineering: Moral Considerability and the Convergence Hypothesis: The Ethics of Geoengineering," *Journal of Applied Philosophy* 29, no. 3 (August 2012): 243–56, https://doi.org/10.1111/j.1468-5930.2012.00568.x.

59. Clare Heyward, "Is There Anything New under the Sun? Exceptionalism, Novelty, and Debating Geoengineering Governance," in *The Ethics of Climate Governance*, ed. Aaron Maltais and Catriona McKinnon (Lanham, MD: Rowman & Littlefield Publishers, 2015).

60. Heyward, 137.

61. Christopher J. Preston, ed., *Engineering the Climate: The Ethics of Solar Radiation Management* (Lanham, MD: Lexington Books, 2012); Christopher J. Preston, ed., *Climate Justice and Geoengineering: Ethics and Policy in the Atmospheric Anthropocene* (London: Rowman & Littlefield International, Ltd., 2016).

62. Christopher J. Preston, "Ethics and Geoengineering: Reviewing the Moral Issues Raised by Solar Radiation Management and Carbon Dioxide Removal: Ethics & Geoengineering," *Wiley Interdisciplinary Reviews: Climate Change* 4, no. 1 (January 2013): 23–37, https://doi.org/10.1002/wcc.198.

63. Toby Svoboda, *The Ethics of Climate Engineering: Solar Radiation Management and Non-Ideal Justice* (New York: Routledge, 2017).

64. For the purposes of foreshadowing, the Proportionality Principle states: decision-making power should be proportional to the claims that individuals have to influence the decision.

Chapter Two

Research

§1 INTRODUCTION

As mentioned in the introductory chapter, the next two chapters of this book focus on ethical and moral issues surrounding research into and deployment of geoengineering.[1] This chapter focuses on four arguments related to *research* into the technology. Most environmental ethicists, myself included, are not enthused about the idea of engineering the climate. However, despite their unease, most of those in the environmental ethics community still think that modest research into geoengineering should continue. That tepid consensus notwithstanding, there are two prominent arguments that are raised against research: the slippery slope argument and the moral hazard argument. This chapter analyzes both of these arguments in detail.

But before turning to the arguments raised *against* research, we'll first look at two arguments that speak *in favor* of research. The idea of "arming the future" with effective geoengineering technology and the thought that solar radiation management is too cheap *not* to investigate are often cited in support of continued research. Sections 2 and 3 critically look at these arguments and what it is that we should conclude from them. Sections 4 and 5 then address the slippery slope argument and the moral hazard argument, respectively. Section 6 concludes the chapter.

§2 ARMING THE FUTURE

You might remember the name Paul Crutzen from the introductory chapter. Crutzen is often credited with bringing the discussion around geoengineering back out of the shadows. In his 2006 article,[2] Crutzen emphatically states that emissions mitigation is the best policy option for avoiding anthropogenic climate change. However, noting the lethargic pace at which policy makers were taking this route, Crutzen advocates for research into solar geoengineering in order to have it on the shelf were we to ever experience "drastic climate heating" sometime in the future.

In essence, Crutzen's argument—which, following Stephen Gardiner, we will call the "arm the future argument"[3] —goes like this:

1. We are not aggressively reducing our emissions.
2. Given that we are not reducing our emissions, there may come a point in the future at which we will have to choose between allowing a climate-related catastrophe, on the one hand, and deploying geoengineering technologies to avoid such harms, on the other.
3. Both allowing a climate-related catastrophe and deploying geoengineering are to be avoided, but geoengineering is the better of the two options.
4. If we are forced to choose, we should choose the "lesser evil." But in order to choose the "lesser evil" of geoengineering in the future, we need to start researching it now.
5. Therefore, we should start to research geoengineering now.

The reasoning behind the arm the future argument seems rather prudent. But Gardiner alleges that such reasoning may be an instance of what he calls "moral corruption." In his fundamental 2011 book, *A Perfect Moral Storm*,[4] he argues that climate change presents us with an ethical problem like none we have ever encountered in human history. Climate change constitutes a tripartite ethical storm that is *global*, *intergenerational*, and *theoretical*. The global storm arises from the fact that climate change has profound transboundary effects, with emissions from one source contributing to harm and destruction elsewhere. The intergenerational storm describes the fact that while the current generation can act so as to benefit or burden future generations, there is no such symmetry the other way around; future generations can do nothing to affect us now. The theoretical storm arises due to the fact that our existing ethical theories developed prior to gigantic problems like climate change, and are unable to adequately evaluate issues of intergenerational justice, international distributive justice, and our relationship to nonhuman nature. These three storms come together and make the prospect of moral corruption all too possible. Gardiner explains: "Most prominently, the perfect storm puts pressure on the very terms in which we discuss the environmental crisis, tempting us to distort our moral sensibilities in order to facilitate the exploitation of our global and intergenerational position."[5]

The thought is that if we were truly serious about our moral obligations to the currently destitute and to future generations, we would immediately start aggressive mitigation and there would be no talk of modifying the climate system. While we should fulfill our moral obligations, Gardiner argues that the perfect moral storm corrupts our moral reasoning and tempts us to look toward easier solutions. Easier solutions, like geoengineering, fall into the category of what Alan Weinberg calls a "technological fix."[6] We must be careful not to simply rationalize our own interests: something that would be all too easy when fulfilling our moral obligation of mitigation is as onerous as it is. It is because of our ability to positively explicate those things in our own interest that we must be wary of the possibility of moral corruption.[7]

Is reasoning like that found in the arm the future argument merely an example of us rationalizing what aligns with our own interest, an example of moral corruption? It is important to reiterate that Crutzen, and others who propose this kind of "lesser evil" argument, think that mitigation is by far the best option.[8] They are not hungry, so to say, to engage in geoengineering. Notwithstanding this stipulation in favor of mitigation, Gardiner sees at least five challenges to the arm the future argument.

§2.1 Against Arming the Future

The first challenge Gardiner raises questions to is the so-called nightmare scenario alluded to in the second premise. How are we to identify when it is that we are faced with this dilemma of either allowing climate-related harms or engaging in geoengineering? Or, furthermore, how can we be certain that this nightmare scenario will even come about? Gardiner thinks that both of these challenges to the idea of the nightmare scenario undermine the arm the future argument, but I'm not so sure. With *medium confidence*,[9] the IPCC states:

> With increasing warming, some physical systems or ecosystems may be at risk of *abrupt* and *irreversible* changes. Risks associated with such tipping points become moderate between 0–1°C additional warming. . . . Risks increase disproportionately as temperature increases between 1–2°C additional warming and become high above 3°C, due to the potential for large and irreversible sea-level rise from ice sheet loss.[10]

With the pledges outlined in the Paris Agreement putting us on track for more than 3°C of warming, it seems as though the risk of a large-scale singular event, like a nightmare scenario, is immanent unless we change course dramatically. Now, Gardiner is right to point out that we may have difficulty in being certain that we are actually in such a situation, but we should not doubt that such a scenario is looming in the future if we continue with business as usual. More will be said about the nightmare scenario at the end of this section.

Gardiner's second challenge to the arm the future argument asks whether or not it presents us with a false dilemma. More specifically, Gardiner claims that our options are not limited to, on the one hand, allowing climate change catastrophes or, on the other hand, engaging in risky ventures of manipulating the planet. Gardiner submits, "Sometimes the best way to plan for an emergency is to prevent its arising."[11] This seems like something Crutzen would agree with. Crutzen and others would agree that our best option would be to prevent the emergency from arising at all. He concludes his 2006 article by saying, "Finally, I repeat: the very best would be if emissions of the greenhouse gases could be reduced so much that the stratospheric sulfur release experiment would not need to take place. Currently, this looks like a pious wish."[12] And, more than a decade later, this still looks like a pious wish. The IPCC predicts that we need to stabilize atmospheric GHG concentrations at 450 parts per million (ppm) in order to have a 66 percent chance of limiting warming to 2°C.[13] In 2017, the U.S. National Oceanic and Atmospheric Administration measured the atmospheric GHG concentration at 493 ppm. So, unless the political landscape surrounding climate change negotiations changes dramatically in the near future, the possibility of preventing serious climatic harms through mitigation alone seems very unlikely.

The third challenge to the arm the future argument concerns us owing more to future generations. This is perhaps the most direct application of moral corruption that Gardiner addresses. We are reminded that the nightmare scenario is something that will mostly likely be affecting future generations of the late twenty-first century or perhaps even the twenty-second or twenty-third centuries. With this fact in mind, "the role of the argument becomes to imply that the responsibility of the current generation is (merely) to aid future generations in choosing the best kind of geoengineering possible."[14] If merely choosing the best kind of geoengineering possible is what the proponents of the arm the future argument have in mind, then it would seem to constitute a kind of moral corruption. But proponents, like Crutzen, seem to agree that the primary responsibility of the current generation is to mitigate. Unfortunately, we are failing to fulfil this collective obligation. So, Gardiner is right to point out that the nightmare scenario will be brought about (in part) by our own collective moral failing, and he claims that "Acknowledging this matters because there seems to be an important moral difference between (on the one hand) preparing for an emergency and (on the other hand) preparing for an emergency that is *to be brought about by one's own moral failure.*"[15]

Gardiner refers to this as an example of "moral schizophrenia."[16] But one might question whether the "we" who are preparing for the emergency are the same group as the "we" who are failing to mitigate. Among currently existing generations, there seem to be multiple camps. There is the camp of Senator James Inhoffe, who infamously stands on the floor of the US Senate, espousing lies about the science of climate change and encouraging the citizenry to continue along a path of high economic growth powered by coal and other fossil fuels. This camp is well-represented among current generations, especially American ones. Then there is the camp of James Hansen, Al Gore, Bun Saluth, Christiana Figueres, and others who emphatically support mitigation efforts and want nothing more than a comprehensive treaty capable of stabilizing the climate. There are many of us in the current generation who are urging our governments to meet our collective obligation of protecting "the climate system for the benefit of present and future generations."[17] And then there are others of us who are happy to drag our feet, as long as we can still eat our hamburgers and drive our sport utility vehicles. It is clear that there are two different first-person plural pronouns being used here. This distinction should allow those of us who want to meet our obligation of reducing emissions as well as preparing future generations for a world in which that obligation is not met to avoid the charge of moral corruption.

The fourth challenge that Gardiner raises for the arm the future argument is that of political legitimacy. This challenge is somewhat complicated and Gardiner develops an intricate argument that he labels "the stalking horse argument" in order to reach two main conclusions. The first is that "any argument for the permissibility of geoengineering has to explain the political legitimacy of those institutions charged with making the decision to geoengineer."[18] The second (related) conclusion is that "any successful argument for the permissibility of geoengineering must invoke appropriate norms of justice and community."[19] Given that the arm the future argument is (fatally) silent both on politically legitimate institutions of governance and appropriate norms of justice and community, the argument is incomplete. Here, Gardiner seems to have a pertinent concern. Any project with the ability to have such profound effects throughout humanity and across the globe should be subject to oversight by legitimate political institutions that are guided by appropriate norms of justice. But it seems possible to simultaneously advocate for both further geoengineering research *and* legitimate institutional regulation. More will be said about governance in later chapters of the book.

The final challenge that Gardiner raises for the arm the future argument is another allegation of moral corruption. We are reminded that the arm the future argument is advocating merely for modest research into potential avenues toward engineering the climate. However, Gardiner thinks that even a proposal of modest research might be a symptom of moral corruption:

> In essence, we'd be happy to spend a few million dollars on research that our generation will probably not have to bear the risks of implementing, and we'd be even happier to think that in doing so, we were making a morally serious choice in favor of protecting future generations. . . . What makes us think that our preference for "modest geoengineering research only" is not just another manifestation of moral corruption?[20]

It is true that research into geoengineering *in lieu of* serious emissions abatement would be easier for us, but this is not what the arm the future argument prescribes. The argument and its advocates admit that if our current generation were to make a "morally serious choice in favor of protecting future generations,"[21] we would aggressively mitigate our global greenhouse gas emissions.

Notwithstanding this partial acquittal of the arm the future argument, the reliance upon the idea of a "nightmare scenario" is problematic, as was alluded to earlier. We can be nearly certain that our inaction on emissions mitigation will bring about significant climatic harms in the future. But justifying research into geoengineering through an appeal to an "emergency situation" is problematic in that the same kinds of justifications are used by authoritarian governments to justify what would normally be considered unjust policy. This may be why most advocates of research do not at this point appeal to geoengineering as a means of avoiding a large-scale singular event, like rapid ice sheet loss. Rather, most advocates see the role of geoengineering technologies similar to the role of a tool in a toolkit; a toolkit that includes mitigation, adaptation, conservation, and innovation.[22] The idea is that we could use geoengineering to shave off some of the peak harms that are associated with any given level of mitigation and adaptation, as John Shepherd's famous "napkin diagram" illustrates.[23]

Justifying current research by alluding to a "nightmare scenario" may be problematic. But the arm the future argument need not rely upon a nightmare scenario. The argument could instead say that we should "arm the future" so that they can shave off the peak of any temperature trajectory and thus blunt some of the harmful effects of climate change. Justifying research into geoengineering by appealing to it as a silver bullet to be used in an exceptional emergency situation is a mistake. But the thought of "arming the future" with a better understanding of the potential benefits and risks of geoengineering (in the event that "we" may want it as part of a portfolio of responses to climate change) speaks in favor of continued research.

§3 THE "INCREDIBLE ECONOMICS" OF GEOENGINEERING

I mentioned earlier that perhaps two of geoengineering's most attractive qualities are its quick and profound efficacy and its relatively inexpensive price tag. In fact, the price tag associated with full-time deployment is a mere fraction of what it would cost to achieve the same reduction in warming with emissions mitigation alone. This has led economists like Scott Barrett to call the economics of geoengineering "incredible."[24] Barret notes not only that geoengineering would be cheap, but that it would be nearly costless when compared to emissions mitigation.

For instance, the famous economist William Nordhaus argues that offsetting all greenhouse warming through geoengineering would carry with it a price tag of about $8 billion annually.[25] Similarly, Keith argues that, for a program that would halve the rate of warming over the next century, "the total cost of large scale geoengineering would be about one billion dollars a year."[26] And other estimates are in the same ballpark. Now, billions of dollars annually sounds like a lot of money. But again, this is an insignificant cost when compared to the same cooling effect that could be achieved through carbon dioxide removal or through emissions mitigation. And it is certainly significantly cheaper than the costs that are predicted to accompany the impacts of climate change. Keith writes, "Estimates put worldwide monetary costs of climate change impacts in the neighborhood of one trillion dollars per year by mid-century."[27] So Barrett is right to note that geoengineering is relatively cheap in comparison to (1) mitigation and (2) certain CDR technologies, and (3) the harm that climate change is expected to bring about.[28]

Barrett writes, "Geoengineering and emissions mitigation are substitutes."[29] But the comparison of geoengineering to emissions mitigation is extremely problematic.[30] As noted in the previous chapter, there are at least four reasons to reject the claim that achieving a certain temperature reduction through geoengineering and achieving the same temperature reduction through emissions mitigation should not be considered perfect substitutes.

First, avoiding a particular amount of warming by means of SAI will have significantly different climatic effects than avoiding the same amount of warming through emissions mitigation. Increase in average global surface temperature is only one of the many climatic effects that are brought about by an increase in atmospheric concentrations of GHGs. While SAI could conceivably be a good substitute for mitigation if all we were interested in were average surface temperatures, surely climate policy has a much broader focus. For instance, if we were to avoid all anthropogenic warming by emissions mitigation, we would end up with regional precipitation patterns very similar to those of the preindustrial era. However, injecting enough sulfur into the stratosphere to counteract all anthropogenic warming could produce regional precipitation patterns significantly different to those of the preindustrial era. As is often pointed out, injecting the amount of sulfate aerosols needed to offset *all* warming brought on by human emissions could cause serious disruption to the Asian and African monsoons—an effect that has the potential to catastrophically impact the food security of billions of people.[31] Of course, the decision to deploy SAI is not a binary one in which it is either used to offset all warming or not used at all.[32] Rather, the technology could be used to offset any chosen percentage of anthropogenic warming. We could start by injecting enough sulfur to counteract only 5 percent of anthropogenic warming, and then slowly increase efforts to a final point at which

50 percent of all anthropogenic warming is offset. When used for the offsetting of only *half* of all anthropogenic warming, the impact on regional precipitation, and thus food security, appears to be negligible (and even positive in some computer models).[33] Still, more research is needed in order to better predict effects on regional climates. And even after significant research, SAI could never serve as a perfect substitute for mitigation with respect to all aspects of the global climate.

Second, the harm our greenhouse gas emissions are causing goes beyond climate. While SAI can moderate increases in global temperature, it will do nothing to address the problem of ocean acidification. The pH balance of the ocean is being affected by the increased concentration of carbon dioxide in the atmosphere. Insofar as SAI will not affect the concentration of carbon dioxide, it will not halt or reduce the acidification of our oceans. I don't mean to imply that this is a reason not to push forward with research into the technology. Rather, the fact that SAI is far from a panacea should be seen as yet another reason to continue with strong mitigation efforts.

A third reason not to equate SAI with emissions mitigation is that we know a geoengineered climate would carry with it negative side effects that a similar climate obtained through mitigation would not. One of the known side effects associated with SAI is increased air pollution. The reason we would choose to inject sulfates into the stratosphere, as opposed to the troposphere, is that stratospheric aerosols have a longer lifespan. But even if we inject our aerosols up in the stratosphere, the particles will eventually make their way down to Earth's surface, where they will contribute to air pollution and thus respiratory problems.[34] A second known side effect of SAI is ozone depletion. In the final quarter of the twentieth century, it became evident that chlorofluorocarbons (CFCs) and other substances were causing serious harm to our planet's atmosphere, specifically the stratospheric layer of ozone near the poles.[35] The 1985 Vienna Convention and subsequent 1987 Montreal Protocol limited the production and use of these dangerous substances. Atmospheric ozone has been replenishing over the past three decades, and a complete recovery is expected in fifty years or so.[36] One negative consequence of injecting sulfur into the stratosphere is that it will be a hindrance to atmospheric ozone recovery. This is because sulfuric aerosols will hasten the breakdown of the CFCs already in the atmosphere.[37] However, there are three reasons we should, despite the risk of ozone depletion, continue research into SAI. First, due to the complicated atmospheric chemistry involved, we do not know exactly how much any given quantity of sulfates will hinder ozone recovery. With more research we can get a better idea of exactly what the risk amounts to. Second, the risk SAI poses to ozone recovery will depend upon *when* the technology is used. If SAI were to be deployed in the second half of the twenty-first century after the control measures within the Montreal Protocol have had enough time to nearly eliminate the presence of CFCs in the atmosphere, then the sulfates would have much less of an effect on ozone. Third, this worry about ozone is spurring research into so-called smart particles that could replace sulfate aerosols, retaining their beneficial properties and avoiding many of their downfalls (including ozone depletion). For these three reasons, concern about ozone depletion is not sufficient to abandon research on SAI. But the known harmful side effects give us reason to refrain from considering the technology a perfect substitute for mitigation.

The fourth, and perhaps most important, difference between SAI and mitigation is another category of possible side effects. We *know* that SAI will have negative effects on air pollution and we *know* that the technology will hinder ozone recovery. That knowledge notwithstanding, the extent of the deleterious effect on air pollution and ozone recovery is unknown. This makes these side effects fall into the category of "known-unknowns." We *know* of them, but their exact effect is *unknown*. As David Keith cautions, "The largest concern [with SAI] is not the risks we know but rather a sensible fear of the unknown-unknowns." The potential for unknown-unknowns to have significant negative effects on both human and natural systems provides us with a fourth weighty reason not to equate SAI with emissions mitigation.

Despite this sobering critique, the inexpensive price tag attached to SAI is one of the reasons that research into it ought to go forward. Even including the costs that go beyond deployment—the costs associated with air pollution, the hindrance of ozone recovery, etc.—SAI is still comparatively inexpensive. If the goal of climate policy is to reduce the risks anthropogenic climate change poses to human and natural systems, and if geoengineering can help to achieve that goal without us having to divert (too many) precious resources away from other worthy ends, then we should not get bogged down by the fact that the technology is not a perfect substitute for mitigation. We should recognize it for what it is and evaluate its research, development, and deployment as part of a multifaceted climate policy portfolio. Of course, more investigation is needed in order to determine whether it should be part of our climate policy portfolio. Yet, there are some who hold that, regardless of its efficacy and cost, research into the controversial technology shouldn't go forward. The two most often cited arguments against research are the slippery slope argument and the moral hazard argument. It is to those arguments that we now turn.

§4 THE SLIPPERY SLOPE ARGUMENT

The slippery slope argument (SSA) is perhaps the most direct argument against research into stratospheric aerosol injection. Also captured by the ideas of "scientific momentum," "path-dependency," and "lock-in,"[38] the slippery slope argument warns that we shouldn't engage in research—that is, the publication of papers, the running of computer models, field experiments—because research of SAI will inevitably lead to its deployment. The thought is that research sits at the top of a slippery slope, at the bottom of which awaits the full-scale deployment of a morally objectionable technology.

This worry about research into SAI resting at the top of a slippery slope is not new. In the first philosophical article assessing the ethics of climate engineering, Dale Jamieson warns that even modest research into the topic runs the risk of inappropriate development, and he gives two reasons to think so. First, "We seem to have a cultural imperative that says if something can be done it should be done. For whatever reason technologies in this society often seem to develop a life of their own that leads inexorably to their development and deployment." And secondly, a "research program often creates a community of researchers that functions as an interest group promoting the development of the technology that they are investigating."[39] The thought is that because of our cultural imperative to develop technologies that are within our capabilities, and because of the fact that scientists generally want their projects to continue, if we begin research into SAI, it will lead to deployment.

Jamieson's pertinent words of caution made their mark on the climate engineering debate. Several articles and publications since his 1996 piece have mentioned the idea of a slippery slope. The Royal Society report on geoengineering governance claims, "Scientific momentum and technological and political 'lock-in' may increase the potential for research on a particular method to make subsequent deployment more likely, and for reversibility in practice to be difficult even when technically possible."[40] Stephen Gardiner writes, "It is not clear that geoengineering activities can really be limited to scientific research. . . . In our culture, big projects that are started tend to get done. This is partly because people like to justify their sunk costs; but it is also because starting usually creates a set of institutions whose mission it is to promote such projects."[41] And Albert Lin warns that "[e]ven very basic and safe research . . . could be a first step onto a 'slippery slope,' creating momentum and a scientific lobbying constituency for development and eventual deployment."[42] There are many others who caution about the slippery slope upon which the research into SAI supposedly rests,[43] but they share this same basic structure: research leads to institutional momentum, which leads to eventual deployment.

§4.1 Slippery Slope Arguments

Slippery slope arguments are often invoked in the political and social realm by politicians of a precautionary or conservative nature. Philosophers, thinking perhaps that all instances of slippery slope reasoning are fallacious, have tended to view them with suspicion.[44] But we cannot paint all slippery slope arguments with the same brush; while some indeed are instances of fallacious reasoning, others are valid and sound arguments. Wibren van der Burg developed a typology of slippery slope arguments. He categorized them into one of two distinct forms: one being logical (or conceptual) and the other being empirical (or psychological).[45]

The *logical* variant of the slippery slope obtains when it is argued that acceptance of A logically commits one to the acceptance of B. For instance, if one accepts that Americans have a right to assisted suicide, then one is logically committed to accepting that Californians have a right to assisted suicide: the acceptance of the antecedent clause (that *Americans* have a right to assisted suicide) logically or analytically commits one to the acceptance of the consequent clause (that *Californians* have a right to assisted suicide). The *empirical* variant of the slippery slope obtains when it is argued that the effect of accepting A will eventually lead to the acceptance of B, through psychological and/or social processes. If the old adage of "As goes California, so goes the nation" is true, then it could be that granting Californians access to assisted suicide will, through psychological or social processes, lead to assisted suicide for all Americans.

The logical form of a general slippery slope argument can be further broken down into two variants: the "no-principled distinction" variant and the "soritical" variant.[46] The no-principled distinction variant says simply "that there is either no relevant conceptual difference between A and B, or that the justification for A also applies to B, and therefore acceptance of A will logically imply acceptance of B."[47] Applied to the case of SAI, this no-principled distinction variant would claim that there is no relevant conceptual or moral difference between research into geoengineering and deployment of the technology.[48] But this is not the worry undergirding the slippery slope argument against geoengineering research. The worry is not that there is no principled difference between research and deployment.[49] The worry is that research will inevitably lead to deployment, not that research and deployment are not distinct. So perhaps the soritical variant more accurately captures the force of the slippery slope argument.

The soritical variant of the argument grants the genuine and meaningful distinction between A and B. It says that while position A may be permissible or unobjectionable, there is a chain of morally indistinguishable steps that leads from A to B. As van der Burg explains, "There is no relevant difference between A and m, and m and n, . . . y and z, and z and B, and that therefore, allowing A will in the end imply the acceptance of B."[50] Mapping this soritical variant on to the slippery slope argument against climate engineering research, the idea would be that published papers are morally indistinguishable from computer modelling, computer modelling is indistinguishable from laboratory experiments, laboratory experiments are indistinguishable from very small field experiments, and small field experiments are indistinguishable from slightly larger field experiments. In the end (through incremental increases in the size of the experiment) the result is full-scale deployment. There is something to be said for this soritical variant of the argument because this is often how advocates of geoengineering research see the technology proceeding.[51] But this, again, is not the worry behind the slippery slope argument against geoengineering research. The worry is not that there is a chain with links of slightly more pugnacious kinds of research that starts with simple theoretical papers and ends with full-blown deployment of millions of tons of sulfur into the stratosphere. After all, there seem to be significant distinguishing factors from, say, experiments in a laboratory and open-air experiments. Rather, the slippery slope argument against geoengineering research is of the empirical form.

§4.2 The Slippery Slope Argument against Geoengineering Research

The empirical form of the argument, remember, claims that the effect of accepting A will eventually lead to the acceptance of B, through psychological and/or social processes. So, the effect of accepting or permitting research into SAI will be that, as a result of either psychological and/or social processes, we will eventually accept or permit the deployment of the technology. And remember that according to Jamieson, the Royal Society, and Gardiner, it is the social process of institutional momentum that will lead us from research to deployment. In what follows, I want to distinguish between two slippery slope arguments against SAI research: one modest and one decisive. Consider first:

The (Modest) Slippery Slope Argument against SAI Research

1. If we research SAI, it will lead to deployment (via institutional momentum).
2. We have serious moral reasons not to deploy SAI.[52]
3. Therefore, we should abandon research into SAI.

Let's assume that premise 1 is true. I have my doubts about whether or not this is a justified assumption. But these doubts will be explored below, so for now I set them aside. Let's look at the second premise. The second premise strikes me as true. Almost everyone acknowledges that we have serious moral reasons to avoid the deployment of SAI. These serious reasons stem from thoughts about hubris, caution, respect for nature, and distributional, procedural, and governance concerns. But even if one accepts premise 1 as true and accepts premise 2 as true, we should still question the conclusion. This is due to the simple fact that, while we may have serious moral reasons not to deploy SAI (again, a claim that seems true), we need to know whether we have *decisive* moral reasons not to deploy the technology. Often times we face difficult decisions in which there are serious moral reasons speaking both in favor of and against a particular course of action. In order to strengthen the conclusion that research into SAI should be abandoned, we would need to have *decisive* moral reasons not to deploy SAI. This leads us to the decisive argument:

The (Decisive) Slippery Slope Argument against SAI Research

1. If we research SAI, it will lead to deployment (via institutional momentum).
2. We have *decisive* moral reasons not to deploy SAI.
3. Therefore, we should abandon research into SAI.

The first thing to note about the above argument is that, if we accept its two premises, we are strongly committed to accepting the conclusion. But there are two routes one could take when questioning such an argument.[53] The first route is to evaluate the conditional premise: premise 1. One can question whether the antecedent actually leads inexorably to the consequent.[54] The second route would have us grant the conditional, empirical premise and instead scrutinize the normative premise: premise 2.[55] In what follows, I will explore both routes. I'll argue both that the empirical, conditional premise is questionable (i.e., that we have insufficient evidence to be able to confidently determine its truth value), and that the normative premise requires more justification, a justification that (ironically) may require more research into the technology.

§4.2.1 The Empirical, Conditional Premise

Before offering up an analysis, I want to point out that premise 1 is somewhat vague. First, what is meant by research? Are we referring to graduate students writing papers in philosophy journals? Are we referencing postdoctoral researchers running computer models? Are we referring to public and private research laboratories running enclosed experiments? Or is it field trials that is meant? Or perhaps any and all of this should be considered research. Second, what is meant by institutional momentum? Does institutional momentum require the backing of coercive institutions like governments? Or can institutional momentum be merely the fact that a graduate program has a cohort of students researching the topic and that they will continue researching it (at least) until they finish their degrees? It's hard to evaluate the premise without gaining better insight into what "research" and "institutional momentum" denote. Setting this difficulty aside, there are at least two reasons to question whether research will unavoidably lead to deployment.

The first reason to question the claim is the fact that many times, if not most of the time, research into new technologies does *not* lead to the development of these technologies. Rather, the development of the new technology is abandoned for any number of reasons (e.g., lack of profitability, too many associated risks, lack of demand, etc.). Perhaps the clearest example of this can be found in the pharmaceutical industry. There is data analyzing the rate of failure of new chemical entities that are researched by various pharmaceutical companies. "The data indicate that the average success rate . . . is approximately 11%"[56]; or, put another way, a mere one in nine compounds makes it to development. The data contradicts the claim that research into a new technology (or chemical entity) inevitably leads to development. Granted this is only one industry, but there is evidence to suggest that even worse success rates apply to most other industries.[57] Some of the attrition in these new technologies is due to the fact that they would not be profitable, but some of the attrition is due to the fact that these new technologies would not be safe, which leads to their rejection by regulatory agencies (e.g., the Food and Drug Administration in the United States, and the European Medicines Agency in Europe). This points toward the need for a regulatory institution that would oversee climate engineering research and any potential deployment (a topic we'll return to later). But what the data does not show is that research into new technologies generally (or even most of the time) leads to development.

A second reason to doubt the claim that research into SAI will lead to deployment rests on a distinction between "development" and "deployment." Even if we grant the claim that research leads to development, there is still a leap from development to deployment. Nuclear technology provides two different examples. Consider first the case of South Africa. Beginning in the 1960s, South Africa embarked on a research program to develop nuclear weapons. By the early 1980s, a functioning bomb was in hand. What makes South Africa unique is that, after having successfully assembled (developed) a number of a nuclear bombs, the country decided to end its nuclear ambitions, dismantling its bombs in the late 1980s.[58] That is, state investment in research may have led to *development*, but it did not lead to *deployment*. Rather, through political action, development was actually reversed (despite the significant sunk costs). A second example comes from Germany's decision to turn away from nuclear technology as a power source. Nuclear technology created nearly one-third of Germany's electricity production in 2000.[59] This notwithstanding, two different German governments—one led by the Social Democrats and another led by the Conservatives—decided that the country would abandon its nuclear energy program by 2022, again despite the widespread and expensive infrastructure that had been developed across previous decades.[60] Now, admittedly, the Germans had deployed their nuclear technology for a number of years. But the lesson to take away from these examples is that even once technologies are developed, neither their sustained nor their occasional deployment is inevitable. Even if we grant that research into SAI will likely lead to *development* (a claim that requires more backing), whether or not the technology will necessarily be *deployed* is still an open question.

Another example that casts doubt upon research leading necessarily to deployment even after development is that of human cloning. Remember Jamieson's words of caution: "We seem to have a cultural imperative that says if something can be done it should be done. For whatever reason technologies in this society often seem to develop a life of their own that leads inexorably to their development and deployment."[61] In 1997, Dolly the sheep became a household name. Dolly was unremarkable in almost every way, except for the fact that she had been cloned from another adult female sheep through the use of a technology known as "somatic cell nuclear transfer." The implications according to the vast majority of news outlets at the time were clear: human cloning was not far away.[62] Despite this worry, somatic cell nuclear transfer has not led inexorably to the cloning of humans. In fact, while many countries around the world allow for research into somatic cell nuclear transfer, no one has ever actually cloned a human being[63] and the application of somatic cell nuclear transfer to human cloning is actually forbidden by law in many places around the world, including the United States. Now, in referencing human cloning as an example, I should point out that (to my knowledge) no research program has been specifically aimed at producing a clone of an adult human being (though, many use the technology to clone embryos for research). Nonetheless, human cloning seems straightforwardly within our technological grasp, especially after the successful cloning of two crab-eating macaques in 2018.[64] Yet the cultural imperative that Jamieson warns of—the idea that *if something can be done it should be done*—is not exemplified. The fact that we have the technical capability to do so and yet no one is pursuing human cloning casts doubt upon Jamieson's sociological claim.

Because of the two considerations just mentioned—that is, the lack of empirical evidence that research leads to development and the distinction between development and deployment—we have some reason to question the conditional premise. While these considerations don't show that research into SAI will *not* lead to deployment, they do show that we can't be certain that such research *will* lead to deployment. But even if we grant the conditional premise and assume that research will undoubtedly lead to deployment, the decisive slippery slope argument faces another problem. The normative premise requires more justification. That is, it is unclear whether or not we have decisive moral reasons to avoid SAI deployment.

§4.2.2 The Normative Premise

The claim that we have decisive moral reasons not to deploy SAI is in one way perhaps undeniably true. Even a strong supporter of SAI like David Keith would agree that if we were talking about deploying SAI *today*, we probably have decisive moral reasons not to do so.[65] There are simply too many unknowns and too many possible side effects that could have dramatically negative effects on various populations throughout the globe. But no one is entertaining the idea of deploying SAI today. The proposal we are assessing is one in which the technology is researched now, with the thought that it could one day in the future potentially be deployed to combat climate change or buy us time to decarbonize the global economy. The claim that we have decisive moral reasons never to deploy SAI is a complex claim to examine, one that is worthy of its own discussion entirely. And if there were no scenarios under which it would ever be morally permissible to deploy SAI, the case in favor of research would be significantly weakened (if not completely undermined), for there are good reasons not to engage in expensive research simply for the sake of knowledge.[66] Given the complexity of the claim and the number of variables involved, I'll only be able to provide a cursory exploration here. Such a limitation notwithstanding, even a cursory exploration will show that claiming the normative premise to be true will require more justification (ironically, justification that seems to require more research into stratospheric aerosols).

Following many moral philosophers, I think that facts—such as facts about the likely outcomes of SAI deployment—can provide us with (moral) reasons.[67] Sometimes the facts about a situation make the reasons that speak in favor of a particular action very clear. For example, the fact that I have promised to help you with an important project provides me with a reason to do so. But imagine that I have also (absentmindedly) promised another friend to go to the movies at the same time. In this scenario, I have reason to help you and I have reason to go to the movies with my other friend. I seem to be in a bind. "When we must choose between different possible acts," Parfit writes, "our reasons may conflict, and they can differ in what we can call their force, strength, or weight. . . . If our reasons to act in some way are stronger than our reasons to act in any of the other possible ways, these reasons are decisive, and acting in this way is what we have most reason to do."[68] In the example used above, it is clear that, given the importance of your project, I have decisive moral reasons to help you and skip the movies with my other friend. In relation to SAI, the facts about SAI deployment can provide us with reasons to consider and reasons to avoid such deployment. In other words, the facts about SAI deployment can determine whether or not we have decisive moral reasons not to deploy the technology.

Insofar as we are interested in whether or not we have decisive moral reasons not to deploy SAI, we have to consider different kinds of reasons and their strengths. And there would seem to be two kinds of reasons not to deploy the technology: reasons that address intrinsic aspects of the technology; and reasons that address extrinsic aspects of the technology. Reasons that address extrinsic aspects of the technology are considerations grounded in the effects or consequences of deployment. Reasons that address intrinsic aspects of the technology are considerations that count for or against deployment regardless of the effects that such deployment might engender. I'll briefly explore various reasons of each kind below.

There are two powerful reasons commonly cited that count against SAI deployment, two reasons that address intrinsic aspects of the technology. One is the claim that SAI is inherently incompatible with democracy. The second is the claim that SAI deployment will be disrespectful toward nature, or will transgress some natural boundary that ought not to be crossed. We'll look more closely at this second claim in the next chapter. But what about the thought that SAI is inherently incompatible with democracy?

Szerszynski et al. claim that SAI is inherently incompatible with democracy given that, among other things, it will (a) stretch democratic institutions beyond their breaking point, and (b) require centralized, authoritarian control. My coauthors and I argue that the claim that SAI is inherently undemocratic or incompatible with democracy is unfounded.[69] I'll provide a significantly abridged summary here. Both of the above points that buttress the antidemocratic claim are controversial. The idea that SAI will break democratic institutions is sometimes based upon the idea that the technology will distribute benefits and burdens unevenly, creating an environment of "SAI-winners" and "SAI-losers."[70] But all policy, especially climate change policy, distributes benefits and burdens unequally. The claim that SAI will stretch democratic institutions to their breaking point is unsubstantiated. With respect to (b), it may be true that SAI would require centralized decision-making, but centralized decision-making is not incompatible with democracy. Indeed, "Whether a hierarchical or centralized political system qualifies as authoritarian depends on additional factors, such as the impartiality of rules, accountability, transparency, access, modes of participation, and freedom of expression."[71] Furthermore, many scholars of science and technology studies criticize the idea that technologies in general have innate political characteristics and argue instead that how certain technologies are controlled is, at least partly, under human control.[72] It is, at this point, entirely unclear whether SAI will be controlled in a democratic or authoritarian fashion. But what does seem clear is that the technology is not *inherently* undemocratic.[73]

Keep in mind the modesty of the claim being put forward. I am not arguing that SAI will necessarily be controlled in a democratic way. Rather, my coauthors and I argue that the technology is not *inherently* undemocratic. In the next chapter I'll look at other reasons to think the technology might be intrinsically or inherently problematic. But given the ills that climate change is to bring about, and given the *potential* that SAI has to alleviate some of those ills, it seems doubtful that the reasons grounded in intrinsic aspects of the technology can undergird decisive moral reasons never to deploy it. But what about reasons to disfavor deployment that relate to extrinsic aspects of the technology? I mentioned two such extrinsic aspects earlier in the chapter: namely, that SAI would (a) hinder the recovery of atmospheric ozone and (b) cause massive disruptions to the Asian and African monsoons.

Remember that SAI calls for the release of sulfuric aerosols in the stratosphere, and that these sulfuric aerosols will hasten the breakdown of the CFCs already in the atmosphere.[74] This breakdown of CFCs in the atmosphere will slow ozone recovery. But remember also that, due to the complicated atmospheric chemistry involved, we do not know exactly how much any given quantity of sulfates will hinder ozone recovery. With more research we can get a better idea of exactly what the risk amounts to. Consider next the models showing that injecting enough sulfuric aerosols into the atmosphere to counteract all anthropogenic warming will cause serious disruption to the Asian and African monsoons—an effect that has the potential to catastrophically impact the food security of billions of people. And remember that these models relied upon SAI to counteract *all* anthropogenic warming. There are similar models that seem to predict significantly curtailed precipitation effects when SAI is used to offset less warming. It could even be the case that SAI could offset some of the deleterious changes in precipitation caused by anthropogenic climate change. Though, that is far from certain at this point.

I do not intend to imply that hindering ozone recovery or disrupting precipitation patters are not things we should worry about with SAI. And there are a number of other extrinsic reasons we might have to disfavor deployment. Rather, I mean to claim that supposed extrinsic aspects of the technology are not good candidates to make us think that we have decisive moral reasons to abandon research. This is because research is how we come to know about extrinsic aspects of the technology. Whether or not SAI deployment will (necessarily) hinder ozone recovery or whether or not it will (necessarily) cause disruption to the Asian and African monsoons is not currently known. The problem with evaluating deployment of the technology now is that we still just don't know enough to determine whether its consequences can provide us with decisive moral reasons to avoid such deployment. Ironically, perhaps the only way to determine whether we have decisive moral reasons not to deploy the technology is to do more research into stratospheric aerosols. Thus, premise 2—the claim that we have decisive moral reasons not to deploy SAI—requires more justification. With both premise 1 and premise 2 being questionable, this makes the (decisive) slippery slope argument a weak ground for forgoing research.

Of course, showing that the slippery slope argument fails to ground a moratorium on research is not the same as saying that there is no reason to be concerned with research leading to premature deployment. In order to be confident research does not lead to undesirable deployment, we should have a regulatory institution that is recognized as legitimate. More will be said about this in the final section of this chapter and in Chapter 4. But for now, we move on to another argument against research.

§5 THE MORAL HAZARD ARGUMENT

If the slippery slope argument isn't sufficient motivation to abandon research, perhaps the moral hazard argument is. The moral hazard argument is not specific or novel to geoengineering. Moral hazards can arise in varying strengths and in all different kinds of scenarios, and the term has been used regularly since the late nineteenth or early twentieth century, having been adopted by various disciplines.[75] At its root, the concept describes the change in an individual's attitude toward a certain behavior once some of the costs of that behavior are absorbed by others. For example, lacking flood insurance, I may be reluctant to build my house near a beautiful river for fear of losing it during a flood. However, once I have flood insurance and know that the costs of flood damage will be borne by the larger insurance group of which I am a part, I may be willing to take such a risk. This is because the benefits of taking the risk (i.e., the scenic views of the river) will accrue exclusively to me, whereas the cost (the potential damage to my home) is now absorbed by the larger group of all those paying for insurance. The insurance is emboldening me to engage in socially suboptimal behavior—in other words, it is presenting a moral hazard.

The moral hazard concern, as it relates to geoengineering, expresses the worry that we'll see geoengineering as a kind of insurance against climate change, which will then spur us to continue spewing greenhouse gases into the atmosphere in spite of the urgent need for emissions reductions. Indeed, this is exactly how it is described in the Royal Society's 2009 report: "In the context of geoengineering, the risk is that major efforts in geoengineering may lead to a reduction of effort in mitigation and/or adaptation because of a premature conviction that geoengineering has provided 'insurance' against climate change."[76]

Though he first introduced the term into the discussion surrounding geoengineering, David Keith now thinks that "moral hazard" is not the right term to describe this potential phenomenon.[77] Keith (and others) think the idea of "risk compensation" better captures this potential drawback of research into climate engineering. Risk compensation describes an increase in one's willingness to engage in risky behavior once one perceives the risk to have decreased.[78] "A defining example of risk compensation was the observation that driving fatalities decreased less than expected following the introduction of seatbelts, perhaps because belted drivers went slightly faster."[79] The analog to geoengineering is apparent. The thought is that the prospect of gaining technological control over the climate may lead us to perceive the risks posed by anthropogenic climate change as less daunting than they actually are. This perception of decreased climate risk could, in turn, lead us to continue emitting heat-trapping greenhouse gases even more than we otherwise would have. Regardless of whether we describe the phenomenon as a moral hazard or as risk compensation, should such concerns lead us to abandon research into geoengineering?[80]

Ben Hale has pointed out that "there is a good deal of confusion about what, exactly, the unique moral hazard associated with geoengineering entails."[81] For example, the moral hazard might be that research into geoengineering could weaken our resolve to mitigate our greenhouse gas emissions, as Keith has cautioned.[82] Worse yet, the hazard might be that the prospect of geoengineering will cause us to increase our emissions output, as Martin Bunzl has warned.[83] Or maybe, as the Royal Society remarks, it could divert precious resources away from not only mitigation proposals but also funds meant to finance adaptation to climate change.[84]

The problem of ambiguity for the moral hazard argument is certainly not insurmountable. All one needs to do is formulate the argument in such a way as to clearly demarcate it from other interpretations and then determine whether the proposed phenomenon obtains. But even when the moral hazard argument is unambiguously formulated, there are at least five problems that remain.

§5.1 Problems with the Moral Hazard Argument

First, there is the problem of uncertainty, or unconfirmed or unestablished hazard claims. Take, for instance, Keith's previous formulation of the moral hazard: the claim that research into geoengineering could weaken our resolve to mitigate our greenhouse gas emissions. This claim is uncertain in the sense that it is not definitely known; there is an empirical uncertainty about it. Indeed, we should also recognize the possibility of a positive effect from geoengineering research. As the Royal Society report also notes, there is a chance that the serious consideration or prospective deployment of a geoengineering technology could scare us sufficiently into taking mitigation efforts more seriously.[85] A report from the Natural Environment Research Council seemed to find that not only did public resolve to mitigate emissions not weaken with the prospect of geoengineering, but rather there was evidence in support of the claim that resolve for mitigation and adaptation measures would remain unchanged. The group in the NERC study maintained that "it would be both ethically and practically important to link any new climate change solutions to continued mitigation, recognizing that one solution might not be enough to tackle climate change."[86] There have been a handful of these studies to date. However, there is inconclusive evidence to show that individuals will change their behavior for the worse when confronted with the prospect of geoengineering. In fact, some studies even show a change in individual behavior that points toward *more* weight being placed on mitigation efforts.[87]

But we should be warry of placing too much weight on studies like this, which brings me to the second hurdle the moral hazard argument needs to jump if it is to ground a moratorium on research. Each of these studies are asking individuals to report how their behavior (specifically, their willingness to mitigate) would change if they knew that geoengineering was being researched, developed, and might one day be deployed. But, as has been pointed out by various authors, climate change is not caused by individual action.[88] The real concern when it comes to climate change—and thus, geoengineering—is collective action at the national and international level. Thus, in order for the moral hazard argument to speak against research into geoengineering, we would need to know how large collectives (or the policymakers of those collectives) would alter *their* behavior with respect to mitigation in the face of research into climate engineering.

Third, even if it has been shown that our collective behavior regarding mitigation would alter if a climate engineering proposal were to be feasible, it should then be shown that our altered behavior is inappropriate or detrimental in some way. Hale's conclusion seems apt: "What each argument needs is treatment that attends not only to the phenomenon that individual or collective actors will change their behavior in the wake of policy intervention, but some clarification of what is wrong with changing behavior in that particular way."[89] For example, if our increased emissions were used to build energy efficient hospitals and schools in underserved parts of sub-Saharan Africa, it is not immediately clear that such a change in behavior should be unwelcomed or deemed wrong, notwithstanding the increase in greenhouse gas emissions. Mitigating climate change is one among many worthy causes.

Fourth, if it is shown that the altered behavior is, in fact, detrimental, then we should want to know whether the altered behavior is offset by the benefits of the hazard. Imagine that SAI will cause some members of the global community to change their behavior, and for the worse. The possibility of engineering the climate will embolden them to engage in riskier behavior than they otherwise would have. Martin Bunzl writes, "Such moral hazard is a familiar worry, and we don't let it stop us in other areas: Antilock braking systems and airbags may cause some to drive more recklessly, but few would let that argument outweigh the overwhelming benefits of such safety features."[90] Thus, lacking some solid reason to believe *both* that research into geoengineering will provide a perverse incentive to stall mitigation or—worse yet—increase our emissions, *and* that this negative effect outweighs the potential benefits of alternative behavior that may be associated with such proposals, the moral hazard argument lacks the knock-out punch against climate engineering research some assert it to have.

This seems right, and it leads me to my final concern with the moral hazard argument: specifically, the moral hazard reasoning used against SAI is strikingly similar to the arguments raised against research into the subject of adaptation to climate change. The policy debate around climate change was initially focused solely on the need to reduce greenhouse gas emissions and made little to no mention about the prospects of adaptation, the reason for this being at least twofold.[91] There was genuine belief that avoiding climate change altogether was possible through emissions reductions alone. But, more interestingly, it was argued that research into and a commitment to serious adaptation projects would create a similar moral hazard to that of the prospect of climate engineering. The thought was that the possibility of *adapting* to the coming changes in climate may provide less incentive for the global community to directly address the root cause of climate change through *mitigation* measures.[92]

According to Pielke et al., the lifting of the taboo on discussion of and research into adaptation is attributable to three distinct aspects of climate change, two of which I argue similarly apply to the case of SRM. First, they note that even if emissions reductions of the politically unbelievable kind were to start tomorrow, climate change would remain unavoidable. That is, even if we were to abandon talk of adaptation and focus 100 percent of our attention to emissions mitigation, we would fail to avoid harmful climate change. This means that adaptation—and geoengineering—may still be useful. Second, "vulnerability to climate-related impacts on society are increasing for reasons that have nothing to do with greenhouse gas emissions, such as rapid population growth along coasts and in areas with limited water supplies."[93] So, even if emissions were to drop to zero, vulnerability to climate change would continue increasing. This calls for a separate policy response to reduce vulnerability (i.e., through adaptation) or the climatic hazard (i.e., through climate engineering). And finally, "those who will suffer the brunt of climate impacts are now demanding that the international response to climate change focus on increasing resilience of vulnerable societies."[94] There has so far been insufficient engagement with stakeholders in the developing world regarding their concerns about geoengineering.[95] While it is too early to tell, as unsatisfactory progress on mitigation continues to be the norm, it is entirely conceivable that there may be a demand for research into geoengineering from those who are most vulnerable to climate change.

Given how imperfect of a substitute SAI is for emissions mitigation and adaptation, we should be warry of anything that will weaken our resolve to pursue such measures. Those words of caution notwithstanding, proponents of the moral hazard argument will have to go some way toward addressing the five aforementioned hurdles if the idea of moral hazard or risk compensation is to serve as a ground for a moratorium on research into geoengineering.

§6 CONCLUSION

We know that regardless of our emissions trajectory, we will not be able to avoid all extreme climate-related events. Rather, anomalous weather events due to climate change have already begun. Climate change is not a threat looming in the distant future; it is a current reality. SAI may turn out to be of no use in protecting the vulnerable populations of the world against these climatic changes. It may be that the risks of injecting sulfur into the stratosphere greatly exceed any expected benefit that might accompany the short-term reduction of average surface-level temperature. But, on the other hand, SAI might be able to be used as a kind of stop-gap, buying us time to convert to a zero-carbon economy while softening the blow that is sure to come in the next century. The only way to find out for sure is to do the research and find out exactly what those predicted costs and benefits will be.

Now, the objections I raised to the slippery slope argument and moral hazard argument do not end the debate. The conclusion from the previous discussions is not that there is a moral imperative to research geoengineering or even that it is undoubtedly something that is morally permitted. Rather, the exploration has merely provided us with reasons to think that the slippery slope argument and moral hazard argument against climate engineering research may not ground the moratorium some think they do.

Now, the weakness of the slippery slope argument does not mean that we have nothing to worry about with respect to research into SAI. What the slippery slope argument shows is that there is a serious need for regulation of research and definitely any possible future deployment. In this final section, I'll briefly introduce three institutional design features that could aid a regulatory institution in preventing the premature or undesirable deployment of a technology like SAI.

One institutional design feature that could help deter undesirable deployment would be the use of a stage-gate system, a regulatory approach that would require approval to move from one stage of research to the next. For instance, we could require researchers to get regulatory approval before moving from papers to laboratory experiments, laboratory experiments to small field experiments, and from field experiments to deployment.[96] And, depending upon the strength of the possibility that research will unjustifiably lead to deployment (along with the weight of the bad outcomes were the technology to be deployed), the demandingness required for approval could be fine-tuned. For example, imagine that a regulatory institution is set up with a panel that determines whether to allow research to move to the next stage. If the worry that research will lead to undesirable deployment is immense, then perhaps we should require unanimous approval from the decision-making panel in order to permit moving from one stage to the next. On the other hand, if we wanted to relax the requirement, then perhaps merely a two-thirds majority or even a simple majority of decision-makers would be fitting. However the particular institution is setup, the use of stage gates—with more or less demanding requirements to proceed—could help quell some concerns about research leading inexorably to deployment.

Another promising solution comes from investigative reporting and has been briefly proposed by David Keith, Edward Parson, and M. Granger Morgan.[97] The idea is to have two teams conducting research: a blue team researching the potential benefits of low-cost, low-risk implementation strategies and a completely independent red team tasked with finding holes and proposing problems related to the blue team's research. Coupled with rules of transparency regarding the research produced, Keith et al. notice that "such an adversarial approach may increase the quality and utility of information available to future decision-makers."[98]

Finally, and perhaps most importantly, we could (and almost certainly should) build public engagement directly into the regulatory process. There have been numerous calls for public input when it comes to geoengineering research and development.[99] Public engagement can take various forms, from allowing a comment period on any proposed regulation (as is done by the U.S. Environmental Protection Agency),[100] providing decision-makers with summaries from public focus groups, to embedding public bodies into the actual decision-making process itself. Incorporating or even requiring public engagement in the governance of SAI could serve as an important check on institutional insiders who might be too eager about the technology to put the brakes on research and development.

These are just three options that could be explored when it comes to designing an institution to oversee stratospheric aerosols and other geoengineering technologies. There are, of course, many other options, and it is important to note that the aforementioned options could be used in conjunction, along with other precautionary measures, depending upon how grave we judge the risk of undesirable deployment of SAI. For example, we could provide public deliberative bodies with the findings from the red and blue research teams and imbed these bodies in the stage-gate process, giving them a veto that could stop research from moving from one stage gate to the next.

It is not my intention to argue for one or the other here. Rather, I want to signal that there are tools available to a legitimate regulatory institution to guard against the idea of the slippery slope. This discussion of legitimate governance will be taken up in Chapter 4. The conclusion of this chapter is merely that the previously examined arguments fall short of grounding a moratorium on research.

However, plans to go forward with research only make sense if there is a chance that we would want to deploy the technology. If there are decisive and damning arguments against deployment, we need not do research. It is toward such concerns about deployment that we turn now.

NOTES

1. An earlier version of this chapter was published as Callies, Daniel Edward, "The Slippery Slope Argument against Geoengineering Research," *Journal of Applied Philosophy*, 2018. DOI: 10.1111/japp.12345. Reprinted with permission of John Wiley & Sons.
2. Crutzen, "Albedo Enhancement by Stratospheric Sulfur Injections."
3. Gardiner, "Is Arming the Future with Geoengineering Really the Lesser Evil?"
4. Gardiner, *A Perfect Moral Storm*.
5. Gardiner, 8.
6. Weinberg, *Reflections on Big Science* (Cambridge, MA: MIT Press, 1967).
7. Note that the worry that we will "make a Reason for everything one has a mind to do" is not such a worry from a Humean standpoint. Hume is famous for claiming that "reason is, and ought only to be the slave of the passions." See David Hume, *Treatise of Human Nature*, Book III, Section 3.3.
8. Crutzen, "Albedo Enhancement by Stratospheric Sulfur Injections," 211. It should be pointed out that Crutzen himself did not endorse the "lesser evil" rhetoric employed by Gardiner. See Darrel Moellendorf, *The Moral Challenge of Dangerous Climate Change: Values, Poverty, and Policy* (New York: Cambridge University Press, 2014), 192–202.
9. See IPCC reports for a discussion of their confidence levels, which range from *very low, low, medium, high,* to *very high.*
10. Intergovernmental Panel on Climate Change, *IPCC, 2014: Summary for Policymakers. In: Climate Change 2014: Impacts, Adaptation, and Vulnerability. Part A: Global and Sectoral Aspects. Contribution of Working Group II to the Fifth Assessment Report of the Intergovernmental Panel on Climate Change* (emphasis added).
11. Gardiner, "Is Arming the Future with Geoengineering Really the Lesser Evil?," 292.
12. Crutzen, "Albedo Enhancement by Stratospheric Sulfur Injections," 217.
13. Intergovernmental Panel on Climate Change, *Climate Change 2014*.
14. Intergovernmental Panel on Climate Change, *IPCC, 2013: Summary for Policymakers. In: Climate Change 2013: The Physical Science Basis. Contribution of Working Group I to the Fifth Assessment Report of the Intergovernmental Panel on Climate Change.*
15. Gardiner, 293 (original emphasis).
16. Gardiner, *A Perfect Moral Storm*, 368. Presumably, the idea of "moral multiple-personality-disorder" would be more apt than "moral schizophrenia," given what Gardiner is arguing. He gets the moral schizophrenia terminology from Michael Stocker, "The Schizophrenia of Modern Ethical Theories," *Journal of Philosophy* 73, no. 14 (1976): 453–66.
17. United Nations, "United Nations Framework Convention on Climate Change."
18. Gardiner, "Is Arming the Future with Geoengineering Really the Lesser Evil?," 294.
19. Gardiner, 294.
20. Gardiner, "Is Arming the Future with Geoengineering Really the Lesser Evil?," 295.
21. Gardiner, 295.
22. Keith, *A Case for Climate Engineering* (Cambridge, MA: MIT Press, 2013), xix.

23. Shepherd, "Napkin Diagram."
24. Barrett, "The Incredible Economics of Geoengineering."
25. Nordhaus, *Managing the Global Commons* (Cambridge, MA: MIT Press, 1994).
26. Keith, *A Case for Climate Engineering*, 100.
27. Keith, 101.
28. It is important to highlight that these cost estimates are only looking at the cost of actual *deployment*. As the next paragraph shows, there are other costs that are relevant to the assessment of geoengineering.
29. Barrett, "The Incredible Economics of Geoengineering," 46.
30. To his credit, Barrett does note some of this, which makes the previous quote all the more puzzling.
31. Alan Robock, "20 Reasons Why Geoengineering May Be a Bad Idea," *Bulletin of the Atomic Scientists* 64, no. 2 (May 1, 2008): 14–18, https://doi.org/10.2968/064002006.
32. More will be said about this in Chapter 5.
33. David W. Keith, *A Case for Climate Engineering*.
34. More will be said in Chapter 3 about the potentially troubling trade-off between climate-related deaths and air pollution-related deaths.
35. Scott Barrett, *Environment and Statecraft* (Oxford: Oxford University Press, 2005), 223.
36. National Oceanic and Atmospheric Association, "Science—Ozone Basics," accessed April 29, 2016, http://www.ozonelayer.noaa.gov/science/basics.htm.
37. Crutzen, "Albedo Enhancement by Stratospheric Sulfur Injections."
38. Royal Society op. cit., p. 39.
39. Dale Jamieson, "Ethics and intentional climate change," *Climatic Change* 33, 3 (1996): 323–36, at p. 333. In his article, Jamieson is not addressing stratospheric aerosol injection specifically, but rather the broad category of climate engineering in general.
40. Royal Society op. cit., p. 39.
41. Stephen Gardiner, "Is Arming the Future with Geoengineering Really the Lesser Evil?" in *Climate Ethics*, ed. Stephen Gardiner et al. (Oxford: Oxford University Press, 2010): 284–312, at p. 289.
42. Albert C. Lin, "The Missing Pieces of Geoengineering Research Governance," *Minnesota Law Review* 100, no. 6 (2016): 2509–76, at pp. 32–3.
43. Clive Hamilton, "No, we should not just 'at least do the research,'" *Nature* 496 (2013): 139.
44. Whitman, "The Many Guises of the Slippery Slope Argument."
45. van der Burg, "The Slippery Slope Argument," 43.
46. Jefferson, "Slippery Slope Arguments," 672.
47. van der Burg, "The Slippery Slope Argument," 44.
48. It is debatable as to whether this no-principle distinction variant should even be classified as a form of slippery slope reasoning, rather than just an argument from consistency. See Jefferson, "Slippery Slope Arguments."
49. Though, something similar to that claim is sometimes made. See Macnaghten and Szerszynski, "Living the Global Social Experiment: An Analysis of Public Discourse on Solar Radiation Management and Its Implications for Governance," 467.
50. van der Burg, "The Slippery Slope Argument," 43; Williams, "Which Slopes Are Slippery?"
51. Keith and MacMartin, "A Temporary, Moderate and Responsive Scenario for Solar Geoengineering"; Keith, *A Case for Climate Engineering*; There is also an argument made by Clive Hamilton that could be classified as a logical slippery slope argument of the soritical variant. Hamilton argues that the more we research, debate, and discuss climate engineering, the more it will become accepted in the social sphere. He cautions that we will slowly become more and more accustomed to the links on the research chain so that we will eventually find deployment acceptable. See Hamilton, "No, We Should Not Just 'at Least Do the Research.'"
52. This language is taken from Gardiner op. cit., p. 289.
53. J. A. Burgess, "The Great Slippery-Slope Argument," *Journal of Medical Ethics* 19, no. 3 (1993): 169–74; Anneli Jefferson, "Slippery Slope Arguments," *Philosophy Compass* 9, no. 10 (2014): 672–80; van der Burg op. cit.

54. Burgess op. cit.; van der Burg op. cit.
55. Thomas Douglas, "Intertemporal Disagreement and Empirical Slippery Slope Arguments," *Utilitas* 22, no. 2 (2010): 184–97.
56. Ismail Kola and John Landis, "Can the pharmaceutical industry reduce attrition rates?" *Nature Reviews Drug Discovery* 3, no. 8 (2004): 711–16, at p. 711.
57. Greg A. Stevens and James Burley, "3,000 Raw Ideas = 1 Commercial Success!," *Research-Technology Management* 40, no. 3 (1997): 16–27.
58. Jo-Ansie van Wyk, "Atoms, Apartheid, and the Agency: South Africa's Relations with the IAEA, 1957–1995," *Cold War History* 15, no. 3 (2015): 395–416.
59. International Atomic Energy Association, "Country Nuclear Power Profiles" https://cnpp.iaea.org/countryprofiles/Germany/Germany.htm (accessed 2 February 2018).
60. International Atomic Energy Association op. cit.
61. Jamieson op. cit., p.333.
62. Leon Kass and James Q. Wilson, *The Ethics of Human Cloning* (Washington, DC: AEI Press, 1998), p. viii.
63. Bonnie Steinbock, "Moral Status, Moral Value, and Human Embryos: Implications for Stem Cell Research," in *The Oxford Handbook of Bioethics*, ed. B. Steinbock (Oxford ; New York: Oxford University Press, 2007), 419.
64. "Chinese Scientists Clone Monkeys Using Method That Created Dolly the Sheep," *All Things Considered* (National Public Radio, 2018) https://www.npr.org/sections/health-shots/2018/01/24/579925801/chinese-scientists-clone-monkeys-using-method-that-created-dolly-the-sheep.
65. David W. Keith, *A Case for Climate Engineering*, 12–13.
66. Gardiner op. cit., pp. 288–90. This isn't to deny that there are reasons to pursue knowledge for knowledge's sake. It's merely to point out, as Gardiner does, that we only have so many resources to devote to research and should allocate them responsibly.
67. Derek Parfit, *On What Matters* (Oxford: Oxford University Press, 2011); Thomas Scanlon, *Being Realistic about Reasons* (Oxford: Oxford University Press, 2014).
68. Parfit op. cit., p.32.
69. Joshua B. Horton et al., "Solar Geoengineering and Democracy," *Global Environmental Politics* 18, no. 3 (August 2018): 5–24.
70. Bronislaw Szerszynski et al., "Why Solar Radiation Management Geoengineering and Democracy Won't Mix," *Environment and Planning A* 45, no. 12 (2013): 2809–16.
71. Horton et al. op. cit.
72. Bernward Joerges, "Do Politics Have Artefacts?," *Social Studies of Science* 29, no. 3 (1999): 411–31; Jim Johnson, "Mixing Humans with Non-Humans: The Sociology of a Door-Closer," *Social Problems* 35 (1988): 298–310.
73. Horton et al. op. cit.
74. Paul J. Crutzen, "Albedo Enhancement by Stratospheric Sulfur Injections: A Contribution to Resolve a Policy Dilemma?," *Climatic Change* 77, no. 3–4 (2006): 211–20, at p. 20.
75. Dembe and Boden, "Moral Hazard."
76. Royal Society, *Geoengineering the Climate: Science, Governance and Uncertainty*, 37.
77. Keith, *A Case for Climate Engineering*, 128.
78. Reynolds, "A Critical Examination of the Climate Engineering Moral Hazard and Risk Compensation Concern."
79. Keith, *A Case for Climate Engineering*, 129.
80. Given its prevalence in the literature, I'll continue to use the moral hazard terminology.
81. Hale, "The World That Would Have Been: Moral Hazard Arguments Against Geoengineering," 113.
82. Keith, *A Critical Look at Geoengineering against Climate Change*.
83. Bunzl, "An Ethical Assessment of Geoengineering."
84. Royal Society, *Geoengineering the Climate: Science, Governance and Uncertainty*, 4.
85. Royal Society, *Geoengineering the Climate: Science, Governance and Uncertainty*, 39.
86. Natural Environment Research Council, "Experiment Earth? Report on a Public Dialogue on Geoengineering," 2.

87. See, for instance, Mercer, Keith, and Sharp, "Public Understanding of Solar Radiation Management"; Kahan et al., "Geoengineering and the Science Communication Environment."

88. Sinnott-Armstrong, "It's Not My Fault: Global Warming and Individual Moral Obligations."

89. Hale, "The World That Would Have Been: Moral Hazard Arguments Against Geoengineering," 130.

90. Bunzl, "An Ethical Assessment of Geoengineering," 18.

91. Schipper, "Conceptual History of Adaptation in the UNFCCC Process," 83.

92. Kates, "Cautionary Tales," 6.

93. Pielke et al., "Climate Change 2007."

94. Pielke et al.

95. The Solar Radiation Management Governance Initiative is attempting to change that. See SRMGI, "Solar Radiation Management Governance Initiative."

96. Robert G. Cooper, "A Process Model for Industrial New Product Development," *IEEE Transactions on Engineering Management* EM-30, no. 1 (1983): 2–11.

97. David W. Keith, Edward Parson, and M. Granger Morgan, "Research on Global Sun Block Needed Now," *Nature* 463, no. 7280 (2010): 426–27.

98. Keith, Parson, and Morgan op. cit., p. 426.

99. SRMGI, "Solar Radiation Management Governance Initiative"; Simon Nicholson, Sikina Jinnah, and Alexander Gillespie, "Solar Radiation Management: A Proposal for Immediate Polycentric Governance," *Climate Policy* 18, no. 3 (2018): 322–34; Royal Society op. cit.; Steve Rayner et al., "The Oxford Principles," *Climatic Change* 121, no. 3 (2013): 499–512.

100. Environmental Protection Agency, https://www.epa.gov/superfund/public-comment-process.

Chapter Three

Deployment

§1 INTRODUCTION

As I mentioned, most environmental ethicists (myself included) are not enthused about the idea of engineering the climate. However, most still think that research into geoengineering should continue. Yet, the general consensus that research into the technology should continue relies upon the idea that there are morally permissible deployment scenarios. That is, the main reason to engage in geoengineering research is the thought that there may be scenarios in which we would be willing and morally permitted to deploy the technology. However, many have expressed doubts about whether deployment ever would be the ethically responsible choice. This chapter looks at three such doubts.

In section 2 I take a look at the precautionary argument against deployment. The precautionary argument against deployment builds out of the precautionary principle, which says roughly that "when an activity [like geoengineering] raises threats of harm to human health or the environment, precautionary measures should be taken even if some cause and effect relationships are not fully established scientifically."[1] Drawing upon Sunstein's analysis of the precautionary principle, I argue that the prudent or cautious decision regarding geoengineering must be chosen from the risk/risk scenario we currently find ourselves in (the risk of deploying geoengineering versus the risk of allowing harms from climate change to continue). Under this framing, it is unclear which way the precautionary principle cuts with respect to geoengineering. I then examine another decision-rule for situations with uncertain outcomes—the minimax rule. While minimax reasoning would ground a moratorium on geoengineering, I argue that the conditions that make minimax a reasonable decision-rule don't hold when it comes to geoengineering and climate change. Next, in section 3, I look at the idea of respect for nature. The respect for nature argument claims that even if deployment of geoengineering would prove beneficial for mankind and the planet as a whole, such intentional intervention in the climate system necessarily shows a contempt or disrespect for nature. Following Jamieson, I acknowledge that dominating nature is one way of failing to show it respect. However, while geoengineering may amount to us dominating nature, I argue that it may not be so novel in comparison to many practices we think morally permissible. Finally, in section 4, I address a main deontological worry about deploying geoengineering, namely, the playing God argument. The playing God argument relies upon the doctrine of doing and allowing, saying that since deployment will save some but may condemn others to premature death or hardship, the deployment of the technology is impermissible since it amounts to picking winners and losers. I cast doubt upon the absolutist form of the doctrine of doing and allowing at the state (and international) level, and propose that the doctrine of double effect offers a possible justification for the decision to deploy.

§2 PRECAUTION

As has been mentioned, geoengineering boasts some potentially enormous benefits and some potentially disastrous side effects. Being able to weigh the costs and benefits of deployment, while certainly insufficient to provide a conclusive answer to what we should do, would constitute an important part of our decision-making process. However, developing a cost benefit analysis of geoengineering is difficult since the precise probabilities of the payoffs and undesired side effects remain unknown. Not only that, but the exact effects of climate change are also unknown. How are we to progress in the face of such uncertainty about the benefits and burdens of the technology? Enter the precautionary principle. The precautionary principle is an international norm designed to offer policy guidance in situations wrought with uncertainty, and some invoke the principle when arguing against the future deployment of climate engineering.[2]

§2.1 Precaution in International Law

The principle, in some form or another, has been endorsed by various nations, international governing bodies, and treaties. Perhaps one of its first explicit uses can be found in the 1982 United Nations World Charter for Nature. The parties to the charter agreed that "Activities which are likely to pose a significant risk to nature shall be preceded by an exhaustive examination; their proponents shall demonstrate that expected benefits outweigh potential damage to nature, and where potential adverse effects are not fully understood, the activities should not proceed."[3] Fast-forward ten years to 1992 and we get two similar, yet distinct, formulations. The Rio Declaration reads: "Where there are threats of serious or irreversible damage, lack of full scientific certainty shall not be used as a reason for postponing cost-effective measures to prevent environmental degradation."[4] And the Convention on Biological Diversity reads: "Where there is a threat of significant reduction or loss of biological diversity, lack of full scientific certainty should not be used as a reason for postponing measures to avoid or minimize such a threat."[5] Finally, the world's foremost governing body on climate change, the UNFCCC, formulates its version of the precautionary principle in the following way: "Where there are threats of serious or irreversible damage, lack of full scientific certainty should not be used as a reason for postponing such measures."[6]

§2.2 Precaution, Regulation, and Geoengineering

As is clear from the previous examples, there is no *one* precautionary principle. Rather, the precautionary principle connotes something akin to the idea of "better safe than sorry," and it has been instantiated in policy and law in different ways by different governing bodies. While the wording and, to a large extent, the implications of each formulation vary, Neil Manson identifies a pattern underlying each different construction of the principle.[7] There is always, first, a damage condition; then, second, a knowledge condition; and, finally, a remedy or proposed course of action. Looking at the UNFCCC variant of the principle, Manson's three-part structure is clear. "Where there are threats of serious or irreversible damage [1. The damage condition], lack of full scientific certainty [2. The knowledge condition] should not be used as a reason for postponing such measures [3. The remedy]."[8]

While not being empirically demonstrable or an example a priori reasoning, the precautionary principle enjoys significant support among governmental agents as an integral principle for decision-making under conditions of uncertainty, especially within the environmental realm. Despite its international political support, many theorists are critical. Cass Sunstein, for example, claims that the precautionary principle, rather than offering guidance toward the safest alternative in decision-making, cripples the ones making the decision and denies them any option, thus rendering the principle incoherent.[9]

Sunstein is well aware that the principle does not have one formulation and that it can be interpreted many different ways by different people. This being the case, he makes a clear distinction between what he considers to be the strong version and the weak version of the principle. The weak version can be restated to say that, "A lack of decisive evidence of harm should not be a ground for refusing to regulate."[10] Presumably, when applied to the case of climate engineering, all the weak version of the principle would recommend is that any proposal to manipulate the earth's climate be regulated—a recommendation that few would refuse. Admitting that the weak version of the principle is something to which "no reasonable person could object," Sunstein outlines a stronger version that would, in his words, "require a fundamental rethinking of regulatory policy."[11] That stronger version of the principle reads:

> When there is a risk of significant health or environmental damage to others or to future generations, and when there is scientific uncertainty as to the nature of that damage or the likelihood of the risk, then decisions should be made so as to prevent such activities from being conducted unless and until scientific evidence shows that the damage will not occur.[12]

It is this stronger version of the principle on which Sunstein focusses his criticism, denouncing it as "paralyzing."

When considering how the principle would guide us in regards to climate change, one might reason that even though there is not complete scientific consensus concerning the exact risks posed to human health and the environment, we should err on the side of caution. But what, precisely, does it mean to err on the side of caution in this case? On the one hand, given the dire predictions associated with anthropogenic climate change, it seems as though this strong version of the precautionary principle would recommend regulating greenhouse gas emissions. However, there are also serious threats to human health associated with the act of regulation, that is, with the act of mitigating our emissions. Reducing global greenhouse gas emissions will have a significant negative impact upon the well-being of many throughout the globe, and the poor may feel those impacts the most given that they need GHG-intensive energy to climb out of poverty.[13] Seeing that the precautionary principle could offer two mutually exclusive answers, Sunstein concludes that this "is the sense in which the precautionary principle . . . is paralyzing: It stands as an obstacle to regulation and non-regulation, and to everything in between."[14]

Indeed, this strong version of precautionary principle seems unable to offer coherent policy guidance with respect to geoengineering as well. One way to interpret the principle claims that since climate engineering poses a risk of significant environmental damage and harm to human health, and since there is uncertainty as to the nature of that damage and the likelihood of it, a decision should be made so as to prevent such manipulation of the climate system until we can be certain that such degradation of the environment and harm to human health will not occur. While this may seem like the proper approach and one that owes a large part of its success to the precautionary principle, another interpretation is conceivable. The principle may also demand that governments *prescribe*—rather than *proscribe*—research into and perhaps the future deployment of such technology so as to avoid the risk of significant damage to human health and the environment threatened by anthropogenic climate change *absent* such a technological intervention. Daniel Bodansky recognizes this possible conclusion. He writes, "The problem is that, in the case of climate engineering, failure to take action could also result in irreversible and catastrophic harm due to global warming, so it is unclear which way the principle cuts."[15] If the precautionary principle is going to place a moratorium on any deployment of climate engineering, we will have to abandon this paralyzing version of the principle and get to something that can offer more coherent policy advice.

Recognizing that there is no consensus about the proper formulation of the precautionary principle, and that the principle seems capable of both sanctioning and vetoing the deployment of geoengineering, Lauren Hartzell-Nichols introduces what she calls the "Catastrophic Precautionary Principle" (CPP). The CPP states:

> Appropriate precautionary measures should be taken against threats of catastrophe where: (a) threats of catastrophe are those in which many millions of people could suffer severely harmful outcomes, (b) a precise probability of threat is not needed to warrant taking precautionary measures so long as the mechanism by which the threat would be realized is well understood and the conditions for the function of the mechanism are accumulating, (c) appropriate precautionary measures must not create further threats of catastrophe.[16]

Hartzell-Nichols concludes that, facing the catastrophe of anthropogenic climate change, the CPP would forbid the deployment of stratospheric aerosol injection (SAI) because such a proposal would run afoul of condition (c) above. But, while the CPP offers clear policy guidance with respect to geoengineering, such clear guidance comes at a cost.

First, even Hartzell-Nichols notes that the CPP sets the burden of proof incredibly high. If we held all new technologies to the test that they must prove, beyond a reasonable doubt, that there is no threat of serious harm even in the most unlikely of scenarios, then many things we currently rely upon—airplanes, antibiotics, chlorine, vaccines, radio, X-rays, etc.—all would have been abandoned prior to research even beginning. Secondly, not all catastrophes are the same. In requiring that precautionary measures not carry any threat of catastrophe themselves, we may be closing the door on precautionary measures that, far from being perfect, are a better option than the catastrophe being addressed. The CPP seems to imply a kind of status quo bias in favor of whatever catastrophe first materializes. Ideally, what we need is a principle suitable for situations of uncertainty, but one that takes the costs and benefits of all courses of action into consideration.

§2.3 Minimax Reasoning and the Rawlsian Core Precautionary Principle

Advocating in favor of aggressive mitigation to reduce the risks associated with climate change, Darrel Moellendorf relies upon *minimax* reasoning. Minimax is a decision-rule to be used under conditions of uncertainty, but it avoids some of the pitfalls of Hartzell-Nichols' CPP. The rule advises us to choose the course of action that has the minimum maximum loss, hence the name *minimax*. Put another way, "The rule holds that between courses of action—all with uncertain negative outcomes—the agent should compare only the highest loss scenarios of the courses and choose the course of action that causes the lowest of the highest loss scenarios."[17]

While minimax tells us to minimize our maximum loss, the "maximin" rule (as its name implies) recommends that we maximize our minimum gain.[18] Outside of formal game theory, these two rules offer roughly the same guidance, since minimizing losses and maximizing gains can be seen as two sides of the same coin. Rawls relied upon maximin reasoning when arguing for his two principles of justice.[19] And Stephen Gardiner has applied this reasoning to create what he calls the "Rawlsian Core Precautionary Principle."[20] Very similar to Moellendorf's formulation of the minimax rule, Gardiner's Rawlsian Core Precautionary Principle advises us to assess the possible outcomes of different courses of action, and then decide what to do by choosing the course of action "which has the least bad worst outcome."[21]

Using such reasoning to guide climate change policy, the recommendation is clear. While there is uncertainty regarding both the likely harms to come about from unchecked climate change and from mitigation, the worst-case scenario of unchecked climate change is much worse than the worst-case scenario of mitigation. Moellendorf's minimax reasoning and Gardiner's Rawlsian Core Precautionary Principle both advise us to pursue a policy of aggressive mitigation. But what do these decision-rules imply for the prospect of deploying aerosols in the stratosphere to counteract the effects of climate change? We need to consider the worst possible scenarios of each course of action.

Consider first unchecked climate change. Imagine climate sensitivity is worse than we thought.[22] Imagine that we fail to mitigate and that by the year 2100 we will have brought about a world that is, on average, 5°C warmer than the preindustrial era. This scenario (which, for the sake of argument, assume is the worst-case scenario for unchecked climate change) would represent significant losses for both mankind and nature alike. Consider now geoengineering. The worst-case scenario for the deployment of geoengineering may be the worry mentioned in the previous chapter: that of termination shock. If we were to be headed for a warming of 5°C by 2200 and were to use SAI to mask that warming, then an abrupt termination of deployment would result in a rapid warming of the planet. Now, the main worry with respect to global warming is not the overall increase in temperature (though that is, of course, problematic). Rather, the main worry with global warming is the *rate* at which that increase in temperature occurs. If we were to abruptly stop deployment of geoengineering, we would experience decades of warming in the time span of a year or two. This would be significantly worse than the worst-case scenario of "mere" unchecked climate change. Thus, minimax reasoning and the Rawlsian Core Precautionary Principle would advise against deployment.

But we need to ask, is minimax or the Rawlsian Core Precautionary Principle an appropriate decision-rule for the situation just outlined? Importantly, neither Moellendorf nor Gardiner (nor Rawls, for that matter) think that minimax or maximin reasoning are rational under all circumstances. Indeed, as Moellendorf notes, "It would be implausible to claim that under conditions of uncertainty rational agents should *always* seek the path whose consequence is the lowest maximum negative payout."[23] Rather, there are conditions under which these are plausible decision-rules. Those conditions are:

(a) uncertainty: decision-makers either lack or have reason to discount the probabilities associated with possible outcomes;
(b) potential gains: decision-makers care relatively little for gains that can be made above the highest minimum payout;
(c) acceptable minimum: the rejected alternatives have payouts that are unacceptable, whereas the chosen alternative has an acceptable minimum.

It is not clear whether these are necessary conditions, sufficient conditions, or something else.[24] But what does seem clear, is that neither (b) nor (c) obtain to any great degree when our options are between unchecked climate change and geoengineering. When faced with a choice between unchecked climate change and the deployment of geoengineering to mask that climate change, we care greatly about gains above the minimum payout, and the worst-case scenario of each alternative is unacceptable. Given that two of the conditions under which it is appropriate to use minimax/maximin reasoning don't obtain, it is doubtful that we should rely upon minimax/maximin reasoning and doubtful that the Rawlsian Core Precautionary Principle will offer appropriate guidance when considering whether or not to deploy geoengineering.

§2.4 Concluding Remarks on Precaution

It is important to note the situation of uncertainty we are in with respect to the catastrophic effects associated both with climate change and geoengineering. But what such a simple analysis fails to consider is our diminishing uncertainty on both sides as time progresses. With each year that passes, our understanding of the complexities of the climate system improve. Likewise, the more research conducted on geoengineering, the more we are able to quantify the risks associated with different deployment scenarios. This means that five, ten, or fifty years in the future, the condition of uncertainty necessary for both minimax/maximin reasoning and the precautionary principle may not hold. Five, ten, or fifty years from now, we may be able to attach more precise probabilities to the negative aspects of climate change and geoengineering alike, thus opening the door to both cost-benefit analyses and more informed value judgments about where our priorities should lie. Thus, unless we are planning on deploying geoengineering now under conditions of significant uncertainty—something no one is advocating[25]—it is unlikely the precautionary principle will serve as a decisive reason against deployment.

§3 RESPECT FOR NATURE

Set the idea of uncertainty aside. Imagine that we could be relatively certain that engineering the climate would, in aggregate, be beneficial for humankind. Could we still have reason to forgo such a climate intervention? One reason often cited stems from the idea of *respect for nature*. The thought is that, even if engineering the climate would reliably allay the climatic harms associated with a warmer world, intentionally manipulating the biosphere is simply not the kind of thing us humans should be engaged in. To wit, intentionally engineering the climate amounts to failing to show nature the proper kind of respect it deserves. This disrespect of nature, it is argued, makes engineering the climate the kind of thing that we shouldn't do, regardless of the effects it may have.[26] And if it is granted that deployment would be objectionable by way of disrespecting nature, then research into such proposals need not go forward. This section looks at such concerns.

The thought that engineering the climate is tantamount to disrespecting nature engenders the question: What does it mean to show proper respect for nature? According to Dale Jamieson, "Respecting nature, like respecting people, can involve many different things. It can involve seeing nature as amoral, as a fierce adversary, as an aesthetic object of a particular kind, as a partner in a valued relationship, and perhaps in other ways."[27] Paul Taylor has an even more robust conception of what it means to respect nature. Taylor sees nature—or the organisms that make up nature—as being teleological, with inherent worth that is due proper respect. This respect implies that we have duties "as stringent as to our fellow humans" and that we should avoid "doing harm to or interfering with the natural status of wild living things."[28] This extreme position requires some analysis. Does an organism having a good of its own logically entail that we have reasons to develop an attitude of respect toward it, an attitude that carries with it duties as stringent as we owe to our fellow human beings?

Moellendorf is critical of this strong claim. He argues, "there is a logical gap between the claim that an organism has a good of its own and the claim that we have any sort of duty towards it."[29] He calls this lack of evidence for the claim that teleology implies a respect that carries duties and obligations "the normative gap." By way of counterexample, Moellendorf argues that *Mycobacterium tuberculosis*—the bacterium that causes the infectious tuberculosis disease—certainly has a teleology or a good of its own, but that no one would argue that we have any duties to help it realize that good. Furthermore, we don't even have any duties *not to interfere* with it achieving its own good. We are well within our rights to thwart its teleological goal. Until the normative gap can be bridged, Taylor's position of according nonhuman nature respect equal to that that is accorded to persons seems dubious.

§3.1 Dominating Nature

While it may be difficult to explain what exactly it means to show respect for nature, the more pertinent question is the opposite one: What does it mean to *not* show proper respect for nature? Jamieson suggests, "Dominating something can be one way of failing to respect it [properly], so it is plausible to say that in virtue of our domination of nature we fail to respect it." He continues, "Even if ICC [intentional climate change, or geoengineering] were successful, it would still have the bad effect of reinforcing human arrogance and the view that the proper human relationship to nature is one of domination."[30] That is, even if we were to successfully engineer the climate, this would ingrain in us the idea that our appropriate relationship to nature is one of master and slave. I disagree slightly here with Jamieson. I don't think that engineering the climate in an attempt to minimize the risks posed by climate change reinforces the idea that our *proper* or *appropriate* relationship to nature is one of domination; rather, it may merely be a *permissible* relationship under certain circumstances.

Furthermore, while "dominating" may seem an apt verb to describe our relationship with nature after the deployment of geoengineering technologies, it is definitely an unusual employment of the term from a philosophical point of view. Philip Pettit claims that "a dominated agent, ultimately, will always have to be an individual person or persons" and that an agent dominates this individual person or persons when, "(1) they have the capacity to interfere, (2) on an arbitrary basis, (3) in certain choices that the other is in a position to make."[31] As far as Pettit is concerned, it is a definitional impossibility for us to dominate nature since: (a) it is not a person; and (b) it cannot make choices. And while Jamieson points out that some are happy to speak of nature as "autonomous" or "self-determining,"[32] this would not be enough to meet the standards of Pettit's analysis.[33]

Still, there is merit to the claim that we are dominating nature in some sense of the word, even if not in the strict philosophical sense. "That's what's been happening here for the past ten thousand years," writes Daniel Quinn in his 1991 novel *Ishmael*. "You've been doing what you damn well please with the world. And of course you mean to go right on doing what you damn well please with it."[34] Quinn's novel of forewarning notes that somewhere around 10,000 years ago, we started domesticating animals, irrigating land, and cultivating crops; the first ways in which we began to "dominate" nature. If we were to conclude that "dominating" nature is never permissible and we were to note that agriculture is one of our oldest and most aggressive ways of "dominating" nature, we would have to conclude that agriculture is impermissible. Should we accept such a conclusion and give up our only currently viable means of subsistence? Not even Jamieson thinks this is required. "Perhaps in general we should be more modest in our manipulation of nature," he writes, "but some human changes of the environment are justified and perhaps even morally required."[35] Feasibly, large-scale agriculture is one of these morally required changes that we engage in—even though it is an instance of us "dominating" nature and is often highly destructive to the natural environment.

If it is conceded that agriculture is not a form of "domination," or at least not an impermissible form, could it still be maintained that geoengineering is? Gardiner writes, "To engage in geoengineering would alter the human relationship to this basic background condition [the climate system] and the relationship between humans subject to that condition."[36] Gardiner's claim, I argue, requires more justification. It should be shown how intentionally manipulating the climate is categorically different from intentionally manipulating animals, plants, rivers, and forests, all of which are forms of "domination" that we appear to be willing to accept.

There are two ways in which geoengineering could be categorically different from other ways in which we "dominate" nature. It could be that the *effects* from geoengineering are categorically different than any other kind of "domination" we engage in, or it could be that the "intentional" aspect of geoengineering is what sets it apart. Neither seems like it can ground a categorical difference. We already (unintentionally) inject somewhere around 50 million tons of sulfur into the lower atmosphere through our everyday activities—which has a significant cooling effect on the planet.[37] SAI would add another million tons or so to that total, an increase of 2 percent. It seems implausible that the effect of increasing our air pollutants by 2 percent could engender a categorical difference between dominating and not dominating nature. Does the fact that this effect is brought about intentionally make the difference categorical? This, too, seems implausible since our damning of rivers, plowing of fields, and deforestation are certainly intentional acts as well.

§3.2 Natural Climate Change

Furthermore, even if geoengineering is categorically different from agriculture, deforestation, and other forms of "domination," this does not imply that we are universally unjustified in deploying it. Imagine the average global temperature increases, the sea-level rise, the increased exposure to floods and droughts, and the rest of the harms associated with the climate change we currently face—and will continue to face into the extended future—are not caused by our interference in nature's natural carbon cycle, but rather by normal variation in the climate system. And imagine that under this scenario of "natural climate change" we would be certain to experience a temperature increase of 4°C by the end of the twenty-first century. With an increase of 4°C above the preindustrial average, the IPCC asserts with medium confidence that there will be a "near-complete loss of the Greenland Ice Sheet," which would cause a mean sea level rise of up to 7 meters, swallowing up many low-lying island-states and causing trillions of dollars of damage to major coastal cities such as Guangzhou, New York, and Mumbai.[38] Under this scenario, would we still think that we should show a "proper respect for nature" and let this natural variation take place? Would we not advocate for research into technology capable of preventing such a change and stabilizing the climate at its current evolution or at the level of some point in the recent past? I imagine most would choose to "disrespect" nature in this scenario and deploy geoengineering technology, thereby saving the climate and the biodiversity it supports.

The natural climate change example—in which the abrupt climate change we are currently facing is natural, rather than anthropogenic—seems to cast some doubt upon the thesis that we must always learn to live alongside nature and not to master or "dominate" it. It does not show that geoengineering is not tantamount to dominating nature, but it does lend support to the conclusion that, even if geoengineering would count as another way in which we dominate nature, it may nonetheless be something that we are morally permitted to explore.

§4 PLAYING GOD

As mentioned early in this chapter, geoengineering has the potential to radically reduce global warming. However, alongside this radical potential to cool the globe lie potentially troubling tradeoffs. The benefits of geoengineering are to be accompanied by negative externalities. While reducing the rise of average global surface temperature via solar radiation management

(SRM) has to the potential to save hundreds of thousands to millions of lives, it could at the same time condemn thousands of others to a premature death. It is this intentional tradeoff of harms that may cause us to think that SRM is not to be used, despite the net reduction in harm.

§4.1 Setup and Assumptions

As in many areas of climate science, there is a great deal of uncertainty surrounding the possible effects and side-effects of SRM. In order to focus our attention on the moral arguments as opposed to the sophisticated and unsettled science, I am going to make the following assumptions. First, let's assume (a) that we know SRM will work; assume that the release of one million tons of sulfuric aerosol into the stratosphere will bring about the cooling effect that has been envisioned. Second, let's assume (b) that this cooling effect will save many lives. To be more precise, say that it will save exactly one million people per year from the otherwise fatal weather events that would occur with an absence of such cooling. Third, let's assume (c) that the aerosols that are injected into the stratosphere will eventually reach the earth's surface, causing fatal respiratory diseases that will result in the premature death of ten thousand individuals per year.[39] Fourth, let's assume (simply for the sake of argument) that (e) there is no overlap between the one million individuals who are to be saved from the deadly weather events and the ten thousand who are to die from the respiratory diseases; that is, assume that the two groups are mutually exclusive. Fifth, assume (d) that there is no morally significant difference between any of the 1,010,000 individuals.[40] Finally, assume (f) there are no other relevant side-effects to consider.[41]

What should we conclude about the moral justifiability of SRM given the aforementioned assumptions? In section 4.2 I will introduce the doctrine of doing and allowing and its role in the playing God argument, an argument that would condemn the use of SRM even given the aforementioned assumptions. After laying out the argument, in section 4.3 I will advance two critiques of the doctrine and its role in the argument. Having cast doubt upon the conclusion of the playing God argument, section 4.4 will then explore the doctrine of double effect as a possible moral justification for the deployment of SRM. Some objections to my analysis and potential responses will be explored in section 4.5 before concluding the analysis in section 4.6 with a word on moral dilemmas.

§4.2 Playing God and Doing vs. Allowing

Now, I should mention that I in no way mean to imply that the previously-mentioned assumptions enjoy the kind of scientific certainty that is associated with the claim that greenhouse gases are the primary cause of global warming. The assumptions are fanciful and far too specific. There are good reasons to have some limited faith in some of these assumptions, but this is not the place to defend them. Our question is: If we grant assumptions (a)–(f), could it ever be morally justifiable to deploy SRM?

A simplistic form of utilitarianism would have a clear response to these assumptions and their implications about the moral justifiability of SRM. All other things being equal, the cost-benefit analysis involved in utilitarian reasoning would permit the use of SRM despite the fact that it would bring about the premature death of ten thousand innocent individuals. The benefit derived from the deployment of SRM (the saving of one million people from weather-related deaths) would outweigh the cost associated with deployment (the ten thousand premature deaths that are expected to occur when the aerosols fall from the stratosphere). In fact, not only does this simplistic utilitarianism permit the use of SRM given the aforementioned assumptions, it seems to mandate it.

But this simple cost-benefit analysis that is embedded in utilitarian reasoning often leads to counterintuitive conclusions. In her now widely-used example, Judith Jarvis Thomson illustrates exactly how counterintuitive this reasoning can be.

> Imagine yourself to be a surgeon, a truly great surgeon. Among other things you do, you transplant organs, and you are such a great surgeon that the organs you transplant always take. At the moment you have five patients who need organs. Two need one lung each, two need a kidney each, and the fifth needs a heart. If they do not get those organs today, they will all die; if you find organs for them today, you can transplant the organs and they will all live. But where to find the lungs, the kidneys, and the heart? The time is almost up when a report is brought to you that a young man who has just come into your clinic for his yearly check-up has exactly the right blood-type, and is in excellent health. Lo, you have a possible donor. All you need do is cut him up and distribute his parts among the five who need them.[42]

It has been argued that the classical utilitarian would see the loss of the one healthy young man's life as an unfortunate, but morally permissible, cost given the benefit of saving the five other individuals.[43] And, as David Morrow writes, this has led many to criticize utilitarianism as being insensitive to certain "moral constraints" or moral principles that proscribe such action.[44]

One such relevant moral principle is the doctrine of doing and allowing (DDA). There are multiple formulations of the DDA, but they all get to the same point: doing harm is worse than allowing harm.[45] Of the different formulations of the DDA, some are what Warren Quinn calls "absolutist" and some are "non-absolutist." "Absolutist forms . . . would simply rule out certain choices (for example, murder or torture) no matter what might be gained from them. Non-absolutist forms would simply demand more offsetting benefits as a minimum justification for choices of one sort than for equally harmful choices of the other sort."[46]

The DDA is not an undisputed moral principle, but it enjoys some broad support and offers a simple explanation for many of our moral intuitions. One of these intuitions is the idea that "we shouldn't play God," or we shouldn't implicate ourselves in decisions about who lives and who dies. This exact intuition was invoked by the British minister for home security, Herbert Morrison, in the course of his cabinet debate with Winston Churchill. During World War II, the Nazi V1 bombs were falling just south of the city center of London. Churchill ordered British double agents to pass along false information to the Nazis, telling them the bombs were hitting exactly in the city center, or better yet, slightly north of the center so that the Nazis would alter future trajectories and aim the bombs even further south to a less densely populated area. As David Edmonds explains, Home Security Minister Morrison "was uneasy at the thought of 'playing God,' [that is] of politicians determining who was to live and who to die."[47] Presumably, the thought is that *intentionally* redirecting the Nazi bombs was morally worse than merely *allowing* them to fall on London, even if redirecting them would save more lives.[48]

If we, like Home Security Minister Morrison, were to apply the DDA in its absolutist form, we would arrive at the conclusion that it is worse to *kill* ten thousand through the deployment of SRM than it is to *allow* one million to die from "natural causes." Consider the following argument:

The Playing God Argument

1. If we do not deploy SRM, then we will allow one million people to die due to extreme weather events.
2. If we deploy SRM, we will save those one million people, but will be unintentionally killing ten thousand other people.
3. DDA: killing is worse than allowing to die.
4. Therefore, we should not deploy SRM; it is better to allow one million to die than to kill ten thousand.

If we grant assumptions (a)–(f) mentioned at the beginning of this section, and grant the authority of the absolutist version of the DDA, the playing God argument condemns the deployment of SRM. Not only does it condemn the deployment of SRM, but it also places a presumptive moratorium on research. For, as Morrow notes, "The wisdom of supporting SRM research depends partly on whether it could ever be morally permissible to use SRM as a form of climate engineering."[49] So, to the extent that the playing God argument generates a strong moral reason against the justifiability of deploying SRM, it also generates a strong moral reason against research.

§4.3 Doubts about the Application of the Doctrine of Doing and Allowing

I think most would agree that we always have strong moral reasons to avoid killing, and the DDA is one way of spelling out this judgment. But I want to highlight two reasons to doubt the role of the DDA in the playing God argument. The first is that in its absolutist form, the DDA is a highly counterintuitive and controversial moral principle, and thus shouldn't be relied upon as the basis for an argument. The second is that, even if we grant some weight to the DDA in evaluating the justifiability of individual actions, its wholesale applicability should be questioned at the state level (or the international level, for that matter).

Consider first the DDA in its absolutist form. The absolutist form of the DDA is highly questionable. There are multiple cases in which our intuitions tell us that harming or infringing the right of one is justifiable given that it will save or protect the rights of a greater number of people. The most classic examples in the literature are the infamous trolley problems.[50] When asked whether a bystander is morally permitted to redirect a runaway trolley from a track on which five innocent people will surely be killed to a track on which one innocent person will surely be killed, 90 percent of respondents say that it is permissible for the bystander to redirect the trolley. "Moreover, the judgments appear to be widely shared among demographically diverse populations, including young children; even in large cross-cultural samples, participants' responses to these problems cannot be predicted by variables such as age, sex, race, religion or education."[51] If we were to really take the DDA in its absolutist form seriously, we would have to be willing to judge the action of an individual who diverted the trolley away from, say, one billion people toward one single person as wrong. For most people and for most moral theorists, such a counterintuitive conclusion counts against the absolutist version of the DDA. Another way of putting this is that the absolutist version of the DDA is not a principle that can be endorsed from a position of wide reflective equilibrium.

Furthermore, the absolutist version of the DDA seems even more dubious when it is applied at the state level. Morrow argues that we should view harms resulting from climate change as an instance of "allowing," while we should view harms that are the result of climate engineering as an instance of "doing."[52] But surely both policies should be looked at as roughly on par with one another. On the one hand, we have states actively engaged in policymaking that leads to catastrophic harm due to climate change. On the other hand, we have states actively engaged in policymaking that leads to foreseen harm from climate engineering. What reason do we have to view one policy as doing and one merely as allowing? Consider a parallel example in economic policy. States have various options available to them when it comes to regulating their market economies: toward one end of the spectrum, they can remain rather removed and endorse a more laissez-faire economic policy; toward the other end of the spectrum, they can be thoroughly embedded in the regulation of that market through, for instance, redistributive taxation schemes and the nationalization of certain industries like health care. However, whether states endorse a more unregulated or regulated market, both policies should be considered doings on the part of the state.[53] As another example, consider environmental policy. States have the option of regulating more or less when it comes to harmful pollutants. It may be the case that allowing some pollutants will significantly harm a great number of citizens while delivering benefits to a small corporation. But do we want to say that when the state refrains from passing legislation that would regulate said pollutants that it is merely *allowing* the large group of citizens to be harmed, whereas when it regulates such pollutants it is actively *harming* the small corporation? In contrast, the policy of nonregulation and the policy of regulation should both be seen as active decisions made by the state. Thus, the doing/allowing distinction simply does not seem to apply universally at the state level.

For these two reasons, the absolutist form of the DDA is too controversial to be relied upon as a fundamental premise in an argument. But, tempering what has been said up to this point, the distinction should not be completely thrown out. In fact, most of us still have, as Samuel Scheffler would put it, "a deep commitment to drawing some distinction, in the context of our moral thought, between primary and secondary manifestations of our agency"[54]— that is, we have a commitment to drawing a distinction between things we do and things we allow. But we can maintain a commitment to this distinction without endorsing its absolutist form. Rather than saying that doing harm is always worse than allowing harm, a non-absolutist form of the DDA might say merely that doing harm is *ceteris paribus* worse than allowing harm, meaning that it is harder to justify doing harm than allowing harm (again, *ceteris paribus*). As Morrow construes it, the non-absolutist form of the DDA "says, roughly, that the moral constraint on harming others applies *primarily (or most stringently)* to doing harm to others, and only *sometimes (or less stringently)* to allowing harm to befall others."[55] However, moving to the non-absolutist version of the DDA will then lead to questions about when such a trade-off between doing and allowing meets the standard of justification needed. We turn next to a possible justification of that kind.

§4.4 The Doctrine of Double Effect

In its reliance on the controversial absolutist version of the DDA, the playing God argument is unsound. In fact, not only does the argument seem unsound, my judgment (and the judgments of many others) on the matter directly opposes the conclusion: it seems much better (morally speaking) to save one million people even if this implies condemning ten thousand others, than to refrain from harming the ten thousand thereby allowing one million to die. It isn't just our moral intuitions that back up the thought that saving the greater number is what we ought to do, all things considered. The centuries-old anticonsequentialist doctrine of double effect (DDE) seems to back up this moral intuition and provide us with some guidance about the morality of such trade-offs. The DDE states: "We may not intentionally harm the innocent as an end in itself or as a means to a greater good, but may do a neutral or good act as a means to a greater good even though we foresee that an innocent will be harmed as a side effect."[56]

Originally developed by Saint Thomas Aquinas,[57] Joseph Mangan formulates the principle as follows:

> A person may licitly perform an action that he foresees will produce a good and a bad effect provided that four conditions are verified at one and the same time: (1) that the action in itself from its very object be good or at least indifferent; (2) that the good effect and not the evil effect be intended; (3) that the good effect be not produced by means of the evil effect; (4) that there be a proportionately grave reason for permitting the evil effect.[58]

These four necessary conditions are, according to proponents of the DDE, jointly sufficient to render an act causing both good and evil as morally permissible or justified. Granting assumptions (a)–(f), could SRM deployment meet these four conditions?

The first condition of the DDE says that the object of the action has to be good or at least indifferent. This first condition is meant to rule out acts that are inherently immoral, that is, theft, rape, murder. The action of injecting aerosols into the upper atmosphere is not good, but neither is it bad in and of itself. The same can be said of CO_2 and other greenhouse gases. There is nothing inherently wrong with emitting CO_2. The emission of CO_2 can only be predicated as wrong once it is disruptive to the natural carbon cycle and contributes to the harms that such disruption carries with it. For the vast majority of human existence, the emission of CO_2 was morally indifferent. Likewise, injecting aerosols into the stratosphere is a morally indifferent act by itself. It is the consequences of this act that have moral implications.[59]

The second condition of the DDE, which is the crux of the double effect reasoning, says that the agent must have the good effect, and not the evil effect, as the object of their intention. For instance, a doctor who administers her terminally ill patient a large dose of morphine with the *intention* of hastening the patient's death would be acting impermissibly, whereas if that same doctor were to administer her patient the same dose of morphine with the *intention* of relieving the patient's pain (while merely foreseeing the hastening of death as a side effect), she would be acting permissibly.[60] The intention behind injecting aerosols into the stratosphere is clearly to minimize the effects of climate change, thus saving one million people from weather-related deaths, and not to bring about the ten thousand premature deaths from respiratory diseases. We would still deploy SRM—indeed we would deploy with an even clearer conscience—if such deployment would not bring about any pollution-related deaths. Thus, the death of the ten thousand is merely a foreseen side effect, and not an intended consequence of SRM.

The third condition of the DDE deals with means and ends. It states that the good effect cannot be produced by means of the evil effect. For example, if the death of the ten thousand were causally necessary to bring about the cooling effect needed to save the one million, then even if we did not desire the death of the ten thousand but used their death as an unfortunate means to save the one million, we would be violating the third condition of the DDE.[61] But the antecedent of this hypothetical is counterfactual. The death of the ten thousand is *not* causally necessary to bring about the change needed to save the one million. Thus, the good effect we intend is *not* produced by means of the evil effect.[62]

The fourth and final condition of the DDE addresses proportionality.[63] It requires that there be "a proportionately grave reason for permitting the evil effect."[64] Put another way, the fourth condition asks us to compare our reasons for aiming at the good to our reasons for avoiding the evil. The reasons for aiming at the good have to *significantly* outweigh our reasons for avoiding the evil. If the intended consequence of SRM were to be the saving of one million people, but it were to also have the foreseen side effect of killing 999,999 others, SRM would fail to fulfil this fourth condition of the DDE; the reasons in support of deployment would not be "proportionately grave." And, while sheer numbers play a large role in this calculus of reasons, they are not the only things that matter. For instance, if I can choose between saving the unfortunate occupants of one of two sinking life rafts, and my father sits in one raft while two strangers sit in the other, most would agree that I have more reason to save the raft with my father in it. This is because, while numbers and outcomes certainly generate moral reasons to act, so, too, do obligations and special ties.

So, what about the balance of reasons in our current discussion about the deployment of SRM? We have stipulated that it would save one million people, while condemning ten thousand others. We also stipulated that there are no morally relevant differences between the two groups. Is the one hundred–fold difference of lives saved "proportionately grave" enough to satisfy the fourth condition? No doubt there will be different intuitions on these cases. But if we accept the implausibility of the previously-mentioned absolutist prohibition on causing harm, there is some point at which the significantly greater number of lives saved will tip the balance of reasons in favor of acting. My judgment (and the judgment of most, I would assume) is that saving one million, even when that condemns ten thousand others to a premature death, would be a significant enough difference to tip the balance, thus satisfying the fourth condition. By satisfying all four conditions, the deployment of SRM could be defended by the doctrine of double effect, despite the foreseeable harm it may cause.

§4.5 Objections and Responses

Thus far I've shown why the absolutist version of the doctrine of doing and allowing seems inapt to condemn the deployment of SRM under the conditions outlined in the introduction, and I've introduced the doctrine of double effect as a possible moral justification for the trade-off of harms that will result from using the technology. I now wish to anticipate two objections and outline brief responses to such objections.

The first objection to my analysis may spawn from a confusion about what I see as the potential role for climate engineering. It could be said that if we were truly serious about our moral obligations to the currently destitute and to future generations, we would simply start aggressive mitigation and adaptation measures and there would be no talk of modifying the climate system. But it should be kept in mind that I was not analyzing the moral justifiability of deploying SRM *in lieu of* mitigation and adaption efforts. The question I began with was, given that there will be residual harm despite our best efforts toward mitigation and adaptation, could it ever be morally justifiable to deploy SRM? Mitigation, adaptation, and SRM are not mutually exclusive policy paths: there may be a role for SRM to play in addition to sustained mitigation and adaptation plans.

The second objection one might raise is that my analysis relies upon pure consequentialist reasoning that fails to take deontological constraints seriously. However, I think three considerations undermine the idea that such an analysis is only convincing to consequentialists. First, the doctrine of double effect is not a consequentialist moral principle. To the extent that the second and third conditions of the DDE speak to the "intentions" of the agent being appraised and making sure that the agent is not using the evil as a means to the good end, there are clear moral constraints to applying the doctrine. If the doctrine were an example of pure consequentialist reasoning, there would be no mention of intentions or means. Second, the fourth condition of the DDE requires that there be "a proportionately grave reason" for allowing the evil effect to take place. Again, if this were a purely consequentialist principle, all it would require is that the good effect outweigh the bad effect by the smallest measure of, say, welfare. So, to the extent that the fourth condition of the DDE requires not merely an outweighing of the evil effect but a *proportionately grave outweighing*, the DDE incorporates non-consequentialist reasoning. This brings me to my final point about consequences and moral theory. Taking the good and bad consequences of an action into account when making moral judgments does not automatically limit one's reasoning to the domain of consequentialism. Indeed, contractualist accounts of morality must also pay attention to the benefits and burdens produced by actions.[65] As Scanlon writes, "Any plausible moral view makes what is right or wrong in many cases depend on the harms and benefits to individuals. A theory is consequentialist only if it takes the value of producing the best consequences to be the foundation of morality."[66]

§4.6. Concluding Remarks on the Playing God Argument

I've argued that we have reasons to doubt the absolutist version of the doctrine of doing and allowing and its applicability at the state level. It may be the case that deploying SRM (given the assumptions we've made) requires some kind of moral justification, and I have suggested that the doctrine of double effect is able to provide us with a reason to consider the deployment of the technology under such circumstances as morally justifiable. But this isn't to say that there wouldn't be something morally undesirable about the decision to deploy. While I have argued that deployment may be justifiable and it may be what we ought to do, I have not argued that we should have no pause for concern.

In her exploration of cost-benefit analysis, Martha Nussbaum makes a distinction between what she calls "the obvious question" and "the tragic question." According to Nussbaum, the obvious question is the question we often ask ourselves, the question: What shall we do? "But," she writes, "sometimes we also face, or should face, a different question, which I call 'the tragic question': is any of the alternatives open to us free from serious moral wrongdoing?"[67] When the answer to the tragic question is *No*, we are in an apparent moral dilemma.[68] It means that there are serious moral reasons to avoid each alternative. I think the choice of whether or not to deploy SRM in the scenario we have been examining constitutes an apparent moral dilemma. But that is not to say that it is an irresolvable moral dilemma. The answer to the tragic question may be *No,* in that both alternatives may be axiologically undesirable. But while we have serious moral reasons both to opt for and lobby against deployment, one of the alternatives available to us is significantly better than the other. I argue that, under the aforementioned conditions, our answer to the obvious question of whether or not we should use SRM to reduce the residual harm that is left behind despite our best efforts on mitigation and adaptation could very well be *Yes*.

§5 CONCLUSION

There are few (if any) who think that geoengineering is something we ought to deploy today, or any time soon for that matter. There are simply too many uncertainties and untoward side effects associated with current proposals to warrant serious consideration of immediate deployment. David Keith (one of the technology's strongest supporters) acknowledges that if the choice were between an immediate full deployment of geoengineering, on the one hand, and abandoning the subject of geoengineering forever, on the other, he would choose abandonment.[69] Of course, that is not the decision currently in front of us. In line with the conclusion of the previous chapter, we should continue researching geoengineering in order to find out exactly what its expected costs and benefits amount to.

The arguments surveyed in this chapter, those of precaution, respect for nature, and playing god, fail to show that deployment of geoengineering is something that is categorically morally impermissible. This is most obvious with respect to the precautionary argument. Precaution may advise against immediate deployment of the technology. But as we gain a better understanding of the potentially disastrous effects of anthropogenic climate change, and as our uncertainty about the untoward side effects of geoengineering become quantifiable, the precautionary principle will be unlikely to place an insurmountable moratorium on deployment. It may even cut the other way, prescribing a precautionary approach to climate change that includes all possible policy levers—including geoengineering. Likewise, the idea of respect for nature and the thought that choosing to engineer the climate is, in a way, playing god by deciding who lives and who dies similarly fail to ground a moratorium on future deployment. This is, of course, not to say that engineering the climate, even if done far in the future, is something we should be proud of or something we should welcome. Rather, what the analysis of the aforementioned arguments has shown is that deploying geoengineering is not something that is absolutely morally prohibited.

Still, there is a significant chance that geoengineering could be deployed in morally impermissible ways. And it is very possible that the technology could be used to serve the interests of the powerful, while neglecting the needs and voices of those most vulnerable to climate change. The only way to both pursue the potential benefits of geoengineering while constraining the possibility for it to exacerbate injustice is to ensure legitimate oversight of research, development, and any future deployment. The next three chapters look at what it would take for geoengineering governance to be legitimate, and what it mean to say that that such governance is guided by norms of both substantive and procedural justice.

NOTES

1. Nicholas Ashford et al., "The Wingspread Statement on the Precautionary Principle," January 1998.
2. Lauren Hartzell-Nichols, "Precaution and Solar Radiation Management," *Ethics, Policy & Environment* 15, no. 2 (June 2012): 158–71.
3. United Nations, "World Charter for Nature," 1982, http://www.un.org/documents/ga/res/37/a37r007.htm.
4. United Nations, "Rio Declaration on the Environment and Development," August 1992, http://www.un.org/documents/ga/conf151/aconf15126-1annex1.htm.
5. United Nations, "Convention on Biological Diversity," 1992, http://unfccc.int/essential_background/convention/items/6036.php.
6. United Nations, "United Nations Framework Convention on Climate Change," 1992, http://unfccc.int/essential_background/convention/items/6036.php.
7. Neil A. Manson, "Formulating the Precautionary Principle," *Environmental Ethics* 24, no. 3 (2002): 263–74.
8. United Nations, "United Nations Framework Convention on Climate Change."
9. Cass R. Sunstein, *Laws of Fear: Beyond the Precautionary Principle* (Cambridge: Cambridge University Press, 2005).
10. Ibid., 18.
11. Ibid.
12. Ibid., 19.
13. Ibid., 27.
14. Ibid., 33.
15. Daniel Bodansky, "The Who, What, and Wherefore of Geoengineering Governance," *Climatic Change* 121, no. 3 (December 2013): 542.
16. Hartzell-Nichols, "Precaution and Solar Radiation Management." Hartzell-Nichols adds other conditions to her Catastrophic Precautionary Principle, but these are the most pertinent.
17. Darrel Moellendorf, *The Moral Challenge of Dangerous Climate Change: Values, Poverty, and Policy* (New York: Cambridge University Press, 2014), 81.
18. In formal game theory, minimax and maximin are only the same in zero-sum games. If the games are not zero sum, then minimax refers to minimizing your opponent's maximum gain, while maximin refers to maximizing your own minimum gain. In a non-zero-sum game, these two decision-rules may offer different advice.
19. John Rawls, *Justice as Fairness: A Restatement* (Cambridge, MA: Harvard University Press, 2001), 98.
20. Stephen M. Gardiner, "A Core Precautionary Principle*," *Journal of Political Philosophy* 14, no. 1 (March 1, 2006): 33–60.
21. Ibid., 45.
22. Climate sensitivity is a term that denotes the amount of warming we should expect given a doubling of atmospheric CO_2 concentrations.
23. Moellendorf, *The Moral Challenge of Dangerous Climate Change*, 82 (emphasis added).
24. Moellendorf (83) seems to take these as necessary conditions for relying upon minimax, whereas Gardiner (48) sees them as sufficient conditions that are not strictly necessary. Rawls (99) seems to have seen them as neither necessary nor sufficient, but rather thought that the more of them and the greater degree to which the conditions hold, the more confidence we should have in relying upon minimax or maximin reasoning.
25. David W. Keith, *A Case for Climate Engineering*, Boston Review Books (Cambridge, MA: The MIT Press, 2013), 12–13.
26. Dale Jamieson, "Ethics and Intentional Climate Change," *Climatic Change* 33, no. 3 (July 1, 1996): 323–36; Clive Hamilton, "No, We Should Not Just 'at Least Do the Research,'" *Nature* 496 (April 2013): 139.
27. Dale Jamieson, *Reason in a Dark Time: Why the Struggle Against Climate Change Failed—and What It Means for Our Future* (Oxford: Oxford University Press, 2014), 189.

28. Paul W. Taylor, *Respect for Nature: A Theory of Environmental Ethics*, 25th anniversary ed. (Princeton, NJ: Princeton University Press, 2011), 44.

29. Moellendorf, *The Moral Challenge of Dangerous Climate Change*, 45.

30. Jamieson, "Ethics and Intentional Climate Change," 332. It is important to note, however, that Jamieson admits that geoengineering may be an intrusion or a disrespect of nature that may be morally justified. See below.

31. Philip Pettit, *Republicanism: A Theory of Freedom and Government* (Oxford; New York: Clarendon Press; Oxford University Press, 1997), 52, http://site.ebrary.com/id/10273243.

32. As Jamieson notes, see Eric Katz, *Nature as Subject: Human Obligation and Natural Community*, Studies in Social, Political, and Legal Philosophy (Lanham, MD: Rowman & Littlefield, 1997); Thomas Heyd, ed., *Recognizing the Autonomy of Nature: Theory and Practice* (New York: Columbia University Press, 2005); Jack Turner, *The Abstract Wild*, 3. [ed.] (Tucson: University of Arizona Press, 1999); For the contrary view, see John O'Neill, Alan Holland, and Andrew Light, *Environmental Values* (London: Routledge, 2008).

33. Thanks to Dorothea Gädeke for alerting me to this point.

34. Daniel Quinn, *Ishmael* (New York: Bantam Books, 1995), 161.

35. Jamieson, "Ethics and Intentional Climate Change," 332.

36. Stephen Gardiner, "Is Arming the Future with Geoengineering Really the Lesser Evil?," in *Climate Ethics*, ed. Stephen Gardiner et al. (Oxford: Oxford University Press, 2010), 294.

37. World Health Organization, "Ambient Outdoor Air Quality and Health," accessed August 18, 2017, http://www.who.int/mediacentre/factsheets/fs313/en/.

38. Intergovernmental Panel on Climate Change, *IPCC, 2014: Summary for Policymakers. In: Climate Change 2014: Impacts, Adaptation, and Vulnerability. Part A: Global and Sectoral Aspects. Contribution of Working Group II to the Fifth Assessment Report of the Intergovernmental Panel on Climate Change* (Cambridge: Cambridge University Press, 2014), 13, http://www.ipcc.ch/pdf/assessment-report/ar5/wg2/ar5_wgII_spm_en.pdf. It is important to note that while four degrees of warming would commit us to a near-complete loss of the Greenland Ice Sheet, the sea level rise would occur over a much longer time period (perhaps a millennium).

39. We could also just assume that there would be novel climate configurations that would cause harm.

40. In reality, there may be a morally significant difference. Climate change, as is often reported, will affect developing countries and the poor populations within developed countries much more adversely than it will affect developed nations and the wealthy populations within developing nations. Ambient air pollution, on the other hand, is *somewhat* more evenly distributed across the global population.

41. There will be many, many other effects to consider. Some will be negative effects, such as ozone depletion. But other effects will be positive, such as the preservation of various plant and animal species that are sensitive to sudden temperature increases. We make the assumption that there are no other side effects to consider not because it is a reasonable assumption to make about the way the world will respond to SRM, but because we want to focus our moral inquiry on this one particular aspect of SRM.

42. Judith Jarvis Thomson, "The Trolley Problem," *The Yale Law Journal* 94, no. 6 (May 1985): 1396.

43. This is, of course, a crude interpretation of act-utilitarianism. And even the founding utilitarians have offered accounts of how to respond to just this kind of problem. See Jeremy Bentham, *An Introduction to the Principles of Morals and Legislation* (New York: Barnes & Noble, 2008) Ch. XII Of the Consequences of a Mischievous Act.

44. David R. Morrow, "Starting a Flood to Stop a Fire? Some Moral Constraints on Solar Radiation Management," *Ethics, Policy & Environment* 17, no. 2 (May 4, 2014): 127.

45. For a discussion of the doctrine of doing and allowing, see Philippa Foot, *Virtues and Vices* (Oxford University Press, 2002); Jeff McMahan, "Killing, Letting Die, and Withdrawing Aid," *Ethics* 103, no. 2 (1993): 250–79; Warren S. Quinn, "Actions, Intentions, and Consequences: The Doctrine of Doing and Allowing," *The Philosophical Review* 98, no. 3 (July 1989): 287–312; Judith Jarvis Thomson, "Killing, Letting Die, and the Trolley Problem," ed. Sherwood J. B. Sugden, *Monist* 59, no. 2 (1976): 204–17; Judith Jarvis Thomson, "Turning the

Trolley," *Philosophy & Public Affairs* 36, no. 4 (2008): 359–74; Frances M. Kamm, "Harming Some to Save Others," *Philosophical Studies* 57, no. 3 (1989): 227–60.

46. Quinn, "Actions, Intentions, and Consequences."

47. David Edmonds, *Would You Kill the Fat Man?: The Trolley Problem and What Your Answer Tells Us about Right and Wrong* (Princeton, NJ: Princeton University Press, 2014), 6.

48. Ibid.

49. Morrow, "Starting a Flood to Stop a Fire?," 123.

50. Philippa Foot, "The Problem of Abortion and the Doctrine of Double Effect," in *Virtues and Vices* (Oxford: Oxford University Press, 2002).

51. John Mikhail, "Universal Moral Grammar: Theory, Evidence, and the Future," *Trends in Cognitive Sciences* 11, no. 4 (April 2007): 143–52.

52. Morrow, "Starting a Flood to Stop a Fire?," 130–31.

53. cf. Brian Berkey, "State Action, State Policy, and the Doing/Allowing Distinction," *Ethics, Policy & Environment* 17, no. 2 (May 4, 2014): 147–49.

54. Samuel Scheffler, "Doing and Allowing," *Ethics* 114, no. 2 (January 2004): 218.

55. Morrow, "Starting a Flood to Stop a Fire?," 129 (emphasis added).

56. Kamm, "Harming Some to Save Others."

57. Saint Thomas Aquinas, *Summa Theologica*, 1265, II–II, Qu. 64, Art. 7.

58. Joseph T. Mangan, "An Historical Analysis of the Principle of Double Effect," *Theological Studies* 10, no. 1 (February 1949): 43.

59. This is not to claim that only consequences have import for moral appraisal. Such a statement is consistent with morality being grounded in interpersonal justifiability as well.

60. Alison McIntyre, "Doctrine of Double Effect," in *The Stanford Encyclopedia of Philosophy*, ed. Edward N. Zalta, 2014, http://plato.stanford.edu/archives/win2014/entries/double-effect/.

61. This has been explored in the trolley literature extensively. Imagine you notice the trolley careening down the track toward five innocent people. You are standing on a bridge above the track, and just next to you is a very large man leaning over to see what is happening. You know that you could push the man off the bridge, stop the trolley, and thus save the lives of the five people on the track. Here, the man's death is necessary—his death is the means by which you save the five people (as opposed to the original trolley case where the death of the one individual is merely an unfortunate foreseen side effect of saving the five). This third condition of the DDE would make pushing the large man off the bridge impermissible.

62. This could be expanded and there could be talk about the redundancy of the third condition. See T. A. Cavanaugh, *Double-Effect Reasoning: Doing Good and Avoiding Evil* (Oxford: Oxford University Press, 2006).

63. There is a fifth condition that has been proposed by Michael Walzer. This condition states "that agents minimize the foreseen harm even if this will involve accepting additional risk or foregoing some benefit." See Michael Walzer, *Just and Unjust Wars: A Moral Argument with Historical Illustrations*, Fifth edition (New York: Basic Books, 2015), 151–59.

64. Mangan, "An Historical Analysis of the Principle of Double Effect," 43.

65. Thomas Scanlon, *What We Owe to Each Other* (Cambridge, MA: Belknap Press of Harvard University Press, 1998), 229–31.

66. Thomas Scanlon, "How I Am Not a Kantian," in *On What Matters*, by Derek Parfit (Oxford: Oxford University Press, 2011), 138.

67. Martha C. Nussbaum, "The Costs of Tragedy: Some Moral Limits of Cost-Benefit Analysis," *The Journal of Legal Studies* 29, no. S2 (2000): 1005.

68. Nussbaum thinks that we aren't just in an *apparent* moral dilemma, but that we are in a *genuine* moral dilemma. She ridicules thinkers such as Mill, Kant, and Ross for thinking that moral obligations can "dissolve" given certain circumstances. I disagree with Nussbaum here, and think that the dilemma is merely apparent, and not genuine. But I don't have the space to defend that claim here.

69. Keith, *A Case for Climate Engineering*, 12–13.

Chapter Four

Legitimacy

§1 INTRODUCTION

As we've seen, stratospheric aerosol injection has both promising and troubling characteristics.[1] Consider, for example, its relatively strong leverage on the climate system. The fact that with relatively little effort we could generate huge impacts on global climate gives us reason to look into the technology. However, that same strong leverage makes the technology troubling. In the hands of an irresponsible actor, such technology could have devastatingly negative effects across the world population and could irreparably damage natural ecosystems and the species that comprise them. In a similar vein, the relatively inexpensive price tag attached to the technology means that we could alleviate some future climatic harms without having to divert scarce resources away from mitigation, adaptation, and other valuable goals such as the eradication of global poverty. However, that same relatively inexpensive price tag makes the technology vulnerable to unjustified unilateral action since a multilateral cost-sharing arrangement is not strictly necessary.[2]

Given that such a technology has the potential to bring about both significant benefits and drastic burdens, regulation is important. A legitimate governance institution could ensure that benefits are maximized, that burdens are minimized, that both are distributed in a just manner, and ensure that decisions are being made via justifiable processes. Nearly everyone involved in the normative discussion about climate engineering agrees that *if* we are to move forward with significant research, development, and certainly any future implementation of geoengineering technologies, legitimate governance is a must.[3] Despite this agreement that further research and certainly any future deployment should be accompanied by legitimate governance, there has been relatively little discussion to date among political philosophers about what would constitute legitimate governance of such a technology.[4] That is, while we all might recognize that the abstract *concept* of legitimacy ought to guide geoengineering governance, agreement surrounding the appropriate *conception* of legitimacy for geoengineering governance has yet to emerge.

The main point of this chapter is to introduce a framework out of which an appropriate conception of legitimacy can spring. To do so, I'll begin in the next section by exploring specific conceptions of legitimacy that have been developed in the philosophical literature to date. I'll argue that what we need is a general concept of legitimacy that is appropriate for the diversity of institutions that occupy the international realm—the realm where geoengineering governance would take place. Drawing on the recent work of Allen Buchanan, section 3 then puts forward a general concept of institutional legitimacy. This general concept will pave the way for a more specific normative conception of legitimacy (explored in section 4) that will serve to coordinate judgments about the legitimacy of an institution set up to oversee climate engineering. Section 5 offers a quick recap and then concludes the chapter.

§2 CONCEPTIONS OF (STATE) LEGITIMACY

Before looking at specific *conceptions* of legitimacy, we need to be clear about the general *concept* of legitimacy.[5] In the philosophical literature, the concept of legitimacy has generally denoted either the justification of political power—understood as coercive power backed by government sanctions[6]—or the justification of political authority—understood as a right to rule and a correlating obligation to obey.[7] Aiming at these understandings of the *concept* of legitimacy, various *conceptions* have been developed that spell out exactly when it is that political power or political authority is justified. For example, according to Rawls, "political power is legitimate only when it is exercised in accordance with a constitution the essentials of which all citizens, as reasonable and rational, *can* endorse in the light of their common human reason."[8] Here we see Rawls grounds legitimacy (understood as the justification of coercive power) in the processes whereby political power is exercised.[9] As long as political power is exercised in accordance with a constitution that citizens could hypothetically endorse, then the exercise of such political power is justified. A. J. Simmons has a more demanding conception of legitimacy. For Simmons, a legitimate government has a right to rule and citizens of a legitimate government are under an obligation to obey. The only way, according to Simmons, for a government to derive such a right and for citizens to be under such an obligation to obey is if the citizens have actually expressed their consent.[10] In the absence of such consent, the government has no right to rule and citizens are under no obligation to obey. Given that the legitimacy of the government is conditioned upon voluntary consent, consent theorists like Simmons are called "voluntarists."

Now, the Rawlsian or the voluntarist conception of legitimacy may be an appropriate conception for the legitimacy of state-like institutions.[11] But these conceptions do not apply to international institutions very well. The Rawlsian conception places a priority on the publicly recognized conception of justice embedded in the constitution. And the voluntarist conception places a priority on the express consent of those within the purview of the state. But international institutions exercise political power without any publicly recognized constitution and without the consent of all of those within their purview. Rather than reach the conclusion that international institutions are illegitimate because they fail to meet either the Rawlsian or the voluntarist standards, I contend that we should rely upon a different conception of legitimacy for two reasons.

First, international institutions are beneficial. For example, the world is a better place with the International Atomic Energy Association and the United Nations Framework Convention on Climate Change wielding the power they do, notwithstanding their imperfections. And this is the case despite the fact that they are wielding this power in the absence of any international or global constitution and despite the fact that they do not enjoy the consent of every individual over whom they wield such power. To conclude that these institutions are illegitimate because they fail to meet the standards of state legitimacy outlined by Rawls and Simmons would be counterproductive.[12] But, second, it makes sense to say that how high we set the bar for an institution to be considered legitimate should be sensitive to the characteristics and the function of the particular institution in question. If an international institution is wielding significant power like a state, then perhaps it should meet the demanding requirements we expect of states for them to be considered legitimate. On the other hand, if the institution is merely providing suggestive guidelines and has no ability to enforce any of its directives, then we may want to relax the criteria it needs to fulfil to be considered legitimate.

To be more explicit, what we need is a general concept of institutional legitimacy; a concept that is malleable enough to give rise to appropriate conceptions of legitimacy for the variety of institutions that occupy our world. Clearly explicating a general *concept* of institutional legitimacy will prove invaluable in developing a more concrete *conception* of legitimacy that can be applied to the kind of climate engineering institution that ought to oversee research and possible development.

§3 INSTITUTIONAL LEGITIMACY

To begin the outline of a general concept of institutional legitimacy, we can ask the following question: Why do we have institutions and what is it for an institution, in general, to be legitimate? We have institutions, according to Buchanan, in order solve coordination problems. They supply us with the kind of coordination we need to achieve certain desirable outcomes and do so without the excessive costs or inefficiencies associated with non-institutional alternatives. In order for institutions to effectively coordinate our action and deliver desirable outcomes, they require us to grant them a certain kind of standing—a kind of respect they need to perform their functions.[13] And in order for us to justifiably[14] grant them the respect they need in order to perform this function, we first must converge (or coordinate) on particular normative criteria that institutions ought to fulfil to be worthy of that respect. So, legitimacy assessments serve to solve a metacoordination problem: they allow us to justifiably coordinate around normative criteria that institutions must meet if we are to grant them the respect they need to solve the further problem of coordinating our collective action toward certain desirable outcomes. Given that legitimacy assessments allow us to solve this higher order coordination problem, Buchanan aptly refers to this as the *Metacoordination View* of institutional legitimacy.

The Metacoordination View spells out the general concept of institutional legitimacy. The Metacoordination View says, abstractly, that an institution is legitimate when it is worthy of the respect needed for it to perform its institutional goals. And an institution is worthy of such respect when it sufficiently meets the right normative criteria for the kind of institution it is. This means that a geoengineering governance institution would be legitimate if it is worthy of the respect it needs to coordinate our action around research and development (or abandonment) of technologies capable of modifying the planetary climate. And it would be worthy of such respect when it sufficiently satisfies the right normative criteria. So, the main task in developing a conception of legitimacy for geoengineering governance is specifying the right normative criteria that such an institution ought to fulfil. Before spelling out what I see as the specific normative criteria relevant for the legitimacy of a geoengineering governance institution, I want to highlight some general characteristics of the normative criteria.

§3.1 Characteristics of the Normative Criteria

First, we know that the normative criteria used in legitimacy assessments will vary depending upon the specific form and function of the institution being assessed. The normative criteria appropriate for one kind of institution may be inappropriate for another. Whereas democracy may be a salient normative criterion for state-like institutions, it may not be entirely relevant when making legitimacy assessments of other, non-state institutions. So, the normative criteria proposed here will be specific to geoengineering governance (though, they may apply to other similar institutions as well).

Second, we know that the normative criteria will lie somewhere on a continuum between "the excessively demanding requirements of full justice or optimal efficacy and the excessively forgiving requirement of bare advantage relative to the non-institutional alternative."[15] Moreover, we know that where on this continuum the normative criteria of a given institution fall will depend upon the particulars of the institution and the environment in which the institution is situated. The more important the need for the coordination the institution is providing, the more we should relax our criteria. However, the greater the risk we run by empowering the institution, the more demanding we should make our criteria.[16] The upshot here is that we'll want to make sure the normative criteria identified as appropriate for judging a given institution as legitimate are sensitive to both the particulars of the institution and the environment in which the institution is situated.

Third, we know that the criteria we use to make legitimacy assessments will have to be translated into "epistemically accessible standards" that can serve as proxies for the normative criteria.[17] Imagine that we identify transparency as a normative criterion for geoengineering governance. It may be difficult to agree upon whether or not the abstract normative criterion of transparency is being met. So, in order to coordinate our legitimacy judgments, we could establish a substantive standard relating to the normative criterion of transparency that requires the institution to release its meeting minutes and perhaps its voting record. This would provide a clear point around which addressees of the institution could determine whether or not the criterion of transparency was being fulfilled. The point being made here is that if the abstract normative criteria that ought to be used to coordinate legitimacy assessments are incapable of being translated into substantive standards that agents can use to actually inform their attitudes of respect for the institution, the criteria are inadequate.[18]

Fourth, the proper standpoint from which to determine these normative criteria and judge whether they are being met is social rather than individual. When declaring an institution worthy of respect, we are not declaring that it is worthy of *my* individual respect, in the sense of the first-person *singular* possessive pronoun. Rather, we are declaring that is worthy of *our* respect socially, worthy of our respect in the sense of the first-person *plural* possessive pronoun.[19] This is due to the fact that institutions are attempting to coordinate not my or your action alone, but our action as a group. Whether or not an institution is able to solve the coordination problem it is meant to address will depend upon its success in coordinating the action of the *group* to a sufficient degree. Therefore, to judge a geoengineering governance institution as legitimate means that it is worthy of the respect of the group, not any particular individual within the group.

One final point to mention about the normative criteria is that, according to Buchanan, they are not necessary and/or sufficient conditions for an institution being worthy of our respect.[20] Rather, the more of the criteria that are satisfied, and the greater the extent to which they are satisfied, the more legitimate the institution is. To require that a geoengineering governance institution only be labelled legitimate if it were to fully fulfil all relevant criteria would make the Best the enemy of the Good. Another way of saying this is institutional legitimacy is not a bivalent concept, with institutions either being absolutely legitimate or absolutely not. Rather, it is a concept of degree; an institution can be more, or less, legitimate. Thus, the more and/or the better a geoengineering governance institution fulfils the appropriate normative criteria, the more legitimate the institution is.[21] There will be institutional arrangements that are closer to the end of the spectrum that characterizes clear legitimacy—we can call this *robust legitimacy*. And there will be institutional arrangements that are closer to the end of the spectrum that characterizes clear illegitimacy, while nonetheless sufficiently satisfying the relevant criteria to be worthy of respect—we can call this *weak legitimacy*. Where a geoengineering governance institution falls upon this spectrum will be determined by how well it satisfies the normative criteria that are salient for the kind of institution it is.

§4 NORMATIVE CRITERIA FOR GEOENGINEERING GOVERNANCE

What, then, are the appropriate normative criteria for an institution overseeing geoengineering? There are two methodological obstacles to proceeding with an outline of normative criteria appropriate for a climate engineering regulatory institution. First, without knowing the precise purpose of the institution and the kind of power it has at its disposal to achieve this purpose, nailing down clear and uncontroversial normative criteria for legitimacy assessments is difficult. Not only that, remember that the Metacoordination View sees legitimacy judgments as a *social* practice. The object of the practice is to reach justified agreement on shared normative criteria. Thus, we'll need to decide together what the main goal of this institution is and what kind of power it ought to have to achieve such a goal.

What I offer here, then, are not *the* necessary and sufficient criteria that ought to be used to judge the legitimacy of geoengineering regulation. But given the characteristics of geoengineering technologies, the following proposed criteria seem relevant for assessing whether or not a governance institution is worthy of our respect. Thus, the following criteria are a good place to begin a discourse that will need to gather input from participants from all over the globe. With these words of caution in mind, the following five subsections explore tentative normative criteria for an institution charged with overseeing climate engineering. An institution empowered to oversee climate engineering ought to sufficiently satisfy the normative criteria of: (1) comparative benefit; (2) accountability; (3) transparency; (4) substantive justice; and (5) procedural justice.

§4.1 Comparative Benefit

The first normative criterion of comparative benefit is actually a necessary condition for any justified claim to legitimacy. As mentioned previously,[22] a climate engineering regulatory institution that fails to meet the criterion of comparative benefit ought not to receive our respect. This should be fairly straightforward. Insofar as we are not in a better position to coordinate our action around climate engineering with the institution than we would be without it, we have no reason to accord it the respect it would require to function.

There are a variety of ways in which a geoengineering governance institution could satisfy the comparative benefit criterion relative to the non-institutional alternative. But in order to determine if the institution leads to greater coordination around desirable goals, we'd first need to know what those desirable goals are. This is a decision that is up to the international community, but there are a number of goals that seem reasonable right off the bat. For instance, we'd almost certainly want a governance institution to: (a) help coordinate responsible research and disseminate the results of research to foster a more informed debate about the technology; (b) reduce the threat of unilateral deployment of geoengineering; (c) provide a recognized forum for parties to voice concerns about geoengineering. If a governance institution were to make coordination of responsible research and the dissemination of research more difficult, or were to increase the likelihood of unilateral deployment, or were to hinder the develop of forum for a responsible discussion about geoengineering, then it would fail to fulfil the comparative benefit criterion.[23]

Now, the idea of comparative benefit has two readings. The first is the thought that the institution in question should lead to greater coordination relative to the non-institutional alternative; that is, the world in which we have an institution to oversee climate engineering must enable coordination better than a world in which no such institution exists, *ceteris paribus*. I call this the *non-institutional alternative* reading. The second reading of comparative benefit is also counterfactual, but is not limited to merely a non-institutional alternative. Under this reading of the comparative benefit criterion, the climate engineering regulatory institution must enable greater coordination compared to possible alternative institutions as well. This second—more demanding—reading, which I call the *institutional alternative reading*, has four conditions. A climate engineering institution will fail the comparative benefit criterion if there is an institutional alternative that: (a) provides significantly greater benefits, (b) enjoys similar or even greater feasibility, (c) is accessible without unacceptable transition costs, and (d) sufficiently fulfils the other normative criteria.[24] Thus, a climate engineering institution will be worthy of our respect if it fulfils the comparative benefit criterion under the non-institutional alternative reading. But insofar as there is an alternative institution that satisfies conditions (a)–(d) mentioned above, it is that institution that should be coordinating our action around geoengineering.

§4.2 Accountability

A second normative criterion for the legitimacy of geoengineering governance is the idea of accountability. Remember that empowering an institution to help us solve our coordination problem carries with it risks associated with the power the institution wields. In order to make sure that power is being used in the way we intend it to, we will want institutional agents and the institution itself to be accountable. The norm of accountability has, according to Buchanan and Keohane, three elements:

> First, standards that those who are held accountable are expected to meet; second, information available to accountability holders, who can then apply the standards in question to the performance of those who are held to account; and third, the ability of these accountability holders to impose sanctions—to attach costs to the failure to meet the standards.[25]

To these three I'd like to add a fourth element that would clarify the appropriate group of accountability holders. It wouldn't be hard to imagine a scenario in which the three elements above are properly accounted for, and yet the group of accountability holders is insufficiently restricted so as to render the fulfilment of the first three elements hollow.

While it will be up to the international community to determine the right standards of accountability, we can say broadly that a geoengineering governance institution—indeed, any international institution—must not act outside of the range of powers and prerogatives that are granted to it by the treaty or agreement that brings it into existence. This is what Ngaire Woods refers to as "constitutional accountability." For example, if the institution's remit were confined to promoting responsible research into geoengineering technologies, and the institution were instead funding deployment of the technologies, this would be a clear violation of a standard of accountability. Along with standards of "constitutional accountability," the international community will want to develop standards of what Woods calls "political," "financial," and "internal" accountability as well.[26]

However, while accountability is a normative criterion that a climate engineering regulatory institution ought to satisfy if it is to warrant our respect, it is also possible for such an institution to be overly accountable in at least two ways. First of all, while we want institutional agents to be accountable to the right group, we also want them to exercise their own judgment and expertise to a certain degree. Presumably, institutional agents are better informed than the general group of accountability holders.[27] If institutional agents are to act as responsible representatives for the accountability holders and not merely pander to their wishes, they will require a certain (perhaps small, perhaps larger) degree of insulation.[28] This leads to the second point, which is that—while they certainly should be part of the group of accountability holders—it is impossible to allow future generations to hold current institutional agents accountable. If the current group of accountability holders has too much sway over institutional agents, they can tilt the institution's functioning in their favor, disregarding important duties the current generation may have to future generations. This is what Stephen Gardiner has termed "the tyranny of the contemporary."[29] For these two reasons, we will want to design our accountability standards with some caution.[30]

When it comes to identifying the right constituency of accountability holders, it makes sense to have a tiered approach. Perhaps the most effective accountability holders would be state representatives. State representatives have the political influence to hold international institutions accountable and, at least in democratic states, often do a decent job of representing the interests of their citizens. But given the fact that sometimes states fail to represent the interests of their citizens on the international stage, there will likely be a prominent role for sub-state actors and even collectives of individuals when it comes to holding a geoengineering governance institution accountable.[31]

§4.3 Transparency

A third normative criterion relevant to geoengineering governance is that of transparency. Transparency is commonly defined as "the principle of enabling the public to gain information about the operations and structures of a given entity."[32] While some have highlighted the costs associated with significant transparency,[33] the norm has at least three significant functions. The first significant function correlates to the second element of accountability highlighted above. Transparency aids accountability holders in accessing information useful for determining how they should exercise their power of holding institutional agents accountable. But, secondly, broad transparency gives non-addressees of an institution the ability to analyze the institution's functioning and enables them to "contest the terms of accountability."[34] Finally, transparency can serve as a kind of public justification and create more trust for institutions and the agents that occupy senior positions.[35] And it is worth mentioning, given climate engineering's controversial nature, that transparency will be of the utmost importance in garnering descriptive legitimacy, aiding people in reaching concurrent judgments as to whether or not the institution is worthy of our common respect. However, given the complex and highly technical nature of climate change and especially geoengineering, true transparency will require that the information provided be accessible to both addressees and non-addressees of the institution. This information ought to be accessible not merely in the sense that it is available, but accessible in the sense that it is intelligible to institutional outsiders as well.[36]

When it comes to geoengineering governance, transparency could take many forms. For instance, meeting minutes could be recorded and made public. If the institution is relying upon a voting procedure to make decisions, voting records of the various institutional agents could be released. And when decisions or guidelines regarding, say, field testing of technologies are issued, the institution could release an accompanying document showing the relevant scientific literature upon which its decisions are based. However we incorporate the norm of transparency into governance, we'll want to make sure that it is aiding the public in understanding the rationale behind the processes and decisions coming out of the institution.

§4.4 Substantive Justice

A fourth normative criterion that a climate engineering regulatory institution ought to sufficiently satisfy in order to warrant our respect is captured by the idea of substantive justice. Institutions can deliver certain benefits and often carry with them certain burdens as well.[37] This is certainly true of an institution overseeing geoengineering. It is universally recognized that geoengineering has the potential to create novel distributions of climatic benefits and burdens.

In the abstract, the idea of substantive justice refers to just substantive outcomes, or just distributions of the benefits and burdens produced by a geoengineering governance institution. We generally assume that everyone would like to secure as many of the benefits and as few of the burdens associated with the institution's functioning. We can imagine people having certain claims to some of the benefits and certain claims against the burdens associated with geoengineering governance. Substantive (or distributive) justice obtains when there is a proper balance between the competing claims to those benefits and against the accompanying burdens. While a perfect balance between these competing claims would have to obtain for us to determine the geoengineering governance institution was perfectly just, remember that legitimacy assessments admit of degree. We don't want to require a climate engineering institution to be perfectly just in order for us to grant it the kind of standing it needs to function. This is for at least the following two reasons. First, we may need such an institution in order to make progress on justice. So, "refusing to regard an institution as legitimate unless it is fully just would be self-defeating from the standpoint of justice."[38] Second, requiring that the institution be fully just in order for it to be considered legitimate would conflate "legitimacy" and "justice" and impoverish our moral lexicon. Thus, when making legitimacy assessments, we'll have to determine whether or not a geoengineering governance institution is sufficiently satisfying the demands of substantive distributive justice to be worthy of our collective respect, while nonetheless recognizing that it need not go all the way in satisfying such demands.

Exactly what would count as a just substantive outcome with respect to geoengineering (or climate change in general) is subject to ongoing debate. But there is a modest claim about substantive justice and geoengineering that seems to be on solid ground.[39] The fact that the least well-off members of the global community have (a) contributed the least to the genesis of climate change, (b) have benefited the least from previous actions that have brought about climate change, and (c) have the weakest ability to respond to the burdens of climate change all point to the same conclusion regarding the distribution of benefits and burdens related to geoengineering. The three facts listed above lend support to the following conclusion: if a geoengineering governance institution is to minimally meet the normative criterion of substantive justice, then the distribution of benefits and burdens engendered by such an institution should (probably heavily) favor the least well-off members of the global community. Exactly how much that distribution should favor them is difficult to say.[40] But what seems clear is that if a geoengineering governance institution were to lead to a world in which the least well-off members of the global community were forced to shoulder even greater burdens than they already are in the face of anthropogenic climate change, such an institution would fail to satisfy the normative criterion of substantive justice.[41]

§4.5 Procedural Justice

The final normative criterion I put forward as applicable to geoengineering governance is that of procedural justice. As was noted in the previous subsection, we will want to make sure that our governance institution is conforming to norms of substantive justice. But the substantive distributive outcome with respect to geoengineering is not all we care about. We are also concerned with how it is that we go about making decisions regarding research, development, and deployment. That is, along with substantive justice, we also care about procedural justice. The difference between substantive and procedural justice can be understood as the difference between fairness in the result and fairness in the process, respectively.[42] If we think of substantive justice as a proper balance between the competing claims to the benefits and burdens associated with the institution, we can think of procedural justice as a proper balance between the competing claims to participate in the decision-making process that will determine the outcome. Just as in the case of substantive justice, the normative criterion of procedural justice can be satisfied to a greater or lesser degree. The more procedurally just the geoengineering governance institution is, the more confident we can be that we are justified in granting it the kind of respect it needs to coordinate our action. The more procedural injustice that exists within the institution—that is, the more there are people with claims to participation that are being ignored—the more that will count against us collectively appraising the institution as legitimate.

As in the case of substantive justice, it is difficult to say exactly what would count as a just procedure when it comes to making decisions about geoengineering. But in order for such a procedure to minimally fulfil the normative criterion of procedural justice, it would need to go some way toward providing what I call *fair terms of inclusion* and *fair terms of participation*.[43] By *fair terms of inclusion*, I mean that all those with legitimate claims to participate in the decision-making process are included. By *fair terms of participation*, I mean that all those included in the decision-making process are included on justifiable terms. Again, this doesn't provide us with specific instructions for securing procedural justice in geoengineering governance. But it does help us pick out clear instances of procedural injustice. For example, given the potentially beneficial and potentially catastrophic effects geoengineering could have on, say, those residing within small island states, any process that failed to include these people and allow their voices to be heard on reasonable terms of participation would be procedurally unjust (regardless of the substantive outcome).[44]

Finally, whatever formal procedure ends up guiding the decision-making process around geoengineering, there is a good reason to think that such a procedure should include public participation in some way, shape, or form. From a purely theoretical standpoint, public participation may not be strictly necessary provided that good representatives of all those with legitimate claims are included in the decision-making process. But including public participation in the decision-making process with respect to geoengineering would be valuable on two fronts. First, we know that even the best of representatives do not always fully represent the interests of their constituents—a problem exacerbated at the international level where (a) representatives of democratic states are often too far removed from their constituents to adequately represent their interests, and (b) representatives of nondemocratic states often do not represent the interests of their constituents at all. Including public participation could go some minor way toward addressing this issue. Second, embedding public participation of some kind into the decision-making process can help a geoengineering governance institution secure descriptive or sociological legitimacy more easily.[45] That is, having members of the public included in the institution could help a geoengineering governance institution actually garner the respect it needs to perform its function. And given geoengineering's controversial nature, a governance institution will need all the help it can get in gaining the actual support of the public.

§5. CONCLUSION

A quick recap is probably in order. This chapter began by noting that, given geoengineering's potential to deliver both great benefits and great harms to various populations throughout the world, continued research, development, and any possible deployment ought to be overseen by a legitimate governance institution. And, in order for such an institution to justifiably regulate geoengineering, it will have to be a legitimate institution. While perhaps everyone agrees that geoengineering governance ought to be legitimate, an agreed upon *conception* of legitimacy has yet to emerge. I posited that standard conceptions of (state) legitimacy may be inadequate to determine the legitimacy of a geoengineering regulatory institution. This is because geoengineering governance is something that will take place on the international stage, and international institutions vary dramatically in their characteristics and functions. Given this diversity, we should adopt the broader concept of institutional legitimacy, and then develop specific normative criteria that each kind of institution ought to fulfil. That is the route this chapter has taken, identifying five normative criteria that could be appropriate for making legitimacy assessments of an institution setup to oversee climate engineering.

Now, it's clear that these normative criteria are somewhat ambiguous. And we need to remember that there may be other normative criteria that are also salient to legitimacy assessments of a geoengineering governance institution. Furthermore, even upon reaching a justified agreement regarding the appropriate normative criteria for legitimacy assessments, we still need to explore how these normative criteria can be translated into substantive standards. These standards are to serve as proxies for the normative criteria and, thus, they will need to fall between two points. On the one hand, they need to capture the normative criteria as closely as possible. On the other hand, they need to be accessible to both addressees and non-addressees of the institution, and they need to be standards that could be widely accepted. Exactly how these standards are instantiated and the judgment of whether or not they are met must be done socially.

Some might find this conclusion unsatisfying for perhaps two reasons. First, one might think that geoengineering governance is going to require significant coercive power in order to effectively coordinate our action (which may be the case), and so such a loose conception of legitimacy is insufficient. But remember what was said at the end of section 2. If a geoengineering governance institution is wielding significant coercive power, then it, just like states, ought to meet significantly more demanding normative criteria to be considered legitimate. Second, what we want from a conception of legitimacy, one might argue, are explicit necessary and sufficient conditions that unmistakably spell out exactly when it is that an institution is legitimate and when it lacks legitimacy. The Metacoordination View, clearly, does not do that. But to the extent that we recognize legitimacy assessments as social practices, what the Metacoordination View *does* do is provide us with the right framework from which to engage in that practice. The hope is that this chapter has clearly explicated that general framework, and done some modest work toward generating a conception of legitimacy that will allow us to justifiably coordinate our assessments regarding geoengineering governance. The task of the next two chapters is to provide some content to two specific normative criteria: substantive justice and procedural justice. By spelling out what is meant by substantive justice and procedural justice when it comes to geoengineering, we will have made some headway in crafting specific substantive standards that we can use to either design or evaluate a governance institution.

NOTES

1. An earlier version of this chapter was published as Callies, Daniel Edward, "Institutional Legitimacy and Geoengineering Governance," *Ethics, Policy & Environment*, 2019. DOI: 10.1080/21550085.2018.1562523. Reprinted with permission of Taylor & Francis.

2. For an argument against the prospect of unilateral deployment, see Joshua Horton, "Geoengineering and the Myth of Unilateralism: Pressures and Prospects for International Cooperation," *Stanford Journal of Law, Science & Policy* 4 (2011): 56–69.

3. Note that one need not think that we *should* move forward with research and development to endorse this claim. Even if one thinks research should be abandoned today, this conditional claim could still be accepted. And note that by highlighting the importance of legitimate governance for geoengineering, no claim is being made about its relative importance. There are many other emerging technologies (i.e., gene drives, artificial intelligence, etc.) that deserve attention from normative theorists in order to help guide the development of legitimate regulatory institutions. The focus of this book, however, is on geoengineering. And so it is geoengineering governance that is being considered.

4. For notable exceptions, see Steve Rayner et al., "The Oxford Principles," *Climatic Change* 121, no. 3 (December 2013): 499–512; Pak-Hang Wong, "Consenting to Geoengineering," *Philosophy & Technology* 29, no. 2 (June 2016): 173–88; David R. Morrow, Robert Kopp, and Michael Oppenheimer, "Political Legitimacy in Decisions about Experiments in Solar Radiation Management," in *Climate Change Geoengineering: Philosophical Perspectives, Legal Issues, and Governance Frameworks*, ed. William C. G. Burns and Andrew Strauss (Cambridge: Cambridge University Press, 2013), 146–67; and David Morrow, "International Governance of Climate Engineering: A Survey of Reports on Climate Engineering,

2009–2015," *SSRN Electronic Journal*, 2017. The conception of legitimacy I outline later in the chapter is relatively similar to the conception outlined by Morrow, Kopp, and Oppenheimer—which shouldn't be surprising given that we're all relying upon work done by Allen Buchanan and Robert Keohane. That being said, there are also significant differences. For example, Morrow, Kopp, and Oppenheimer would claim that if a climate engineering regulatory institution is legitimate that it (a) has a right to rule and (b) that those within its purview are under a duty to obey. The Metacoordination View of Buchanan's that I am relying upon claims instead that a legitimate institution is (a) justified in coordinating our action and (b) that those within its purview have a defeasible content-independent reason to comply, but not a duty to obey. Morrow, Kopp, and Oppenheimer also claim that the consent of all democratic states would be necessary but not sufficient to confer legitimacy. But what if all democratic states, except France, consented? Contrary to the view put forward by Morrow, Kopp, and Oppenheimer, the Metacoordination View *could* still consider such an institution legitimate (though this would of course depend upon how well it is satisfying the relevant normative criteria).

5. For explanations of the distinction between concepts and conceptions, see John Rawls, *A Theory of Justice* (Cambridge, MA: Harvard University Press, 1999, p. 5); H. L. A. Hart, *The Concept of Law*, 2nd ed. (Oxford: Oxford University Press, 1998, p. 160); Ronald Dworkin, *Taking Rights Seriously* (London: Bloomsbury, 2013, p. 167); and Christine M. Korsgaard, "Realism and Constructivism in Twentieth-Century Moral Philosophy," *Journal of Philosophical Research* 28 (2003): 99–122.

6. Rawls, *Political Liberalism*, 136.

7. Buchanan, *Justice, Legitimacy, and Self-Determination*, 147.

8. Rawls, *Justice as Fairness*, 41.

9. Though, I should point out that even procedural conceptions like Rawls' recognize substantive constraints. See John Rawls, *Political Liberalism* (New York: Columbia University Press, 2005), 428.

10. Simmons, *Justification and Legitimacy; Essays on Rights and Obligations*, 136.

11. I have serious doubts about the voluntarist conception of legitimacy for even state-like institutions, but I set that concern aside here.

12. I don't mean to suggest that merely being beneficial confers legitimacy onto an institution. Rather, I mean to suggest that we should recognize what we would be giving up by labelling all international institutions as illegitimate. We need international institutions both to make progress on the legitimacy of domestic institutions and on concerns of justice more globally. Thus, to label international institutions as illegitimate would be counterproductive to the causes of making sure that political power is only exercised legitimately and that it serves the cause of justice.

13. Admittedly, "respect" is a complex term with many different variations. Showing respect for a person is certainly distinct from showing respect for, say, the law or another institution. Following Buchanan, we can say that the kind of respect one accords an institution will vary depending upon one's relationship to the institution. In the case of those who are addressees of the institution's directives, showing respect need not mean uniformly complying with those directives or regarding oneself as duty-bound to comply with them. Rather, for those who are addressees of the institution, showing respect implies that institutional directives should be taken seriously, independently of the particular content of the directive. However, for non-addressees, the appropriate form of respect for the legitimate institution will generally equate to at least some kind of presumption in favor of non-interference. For a slightly different understanding of what it means to show an institution the kind of respect it needs to perform its function, see N. P. Adams, "Institutional Legitimacy," *Journal of Political Philosophy* 26, no. 1 (March 2018): 84–102.

14. It should be noted that we are not looking for arbitrary or unjustified granting of this standing. As Buchanan writes, "The goal is not simply agreement that we should support the institution, but rather agreement that is morally appropriate." See Allen Buchanan, *The Heart of Human Rights* (Oxford University Press, 2013), 179.

15. Buchanan, *The Heart of Human Rights*, 193.

16. There are at least two ways that we can be put at risk by empowering an institution. The first, and perhaps most obvious way, is that the institution could abuse its power. For instance,

an institution endowed with significant coercive power could wield that power either through questionable means or to undesirable ends. But, second, we can even be put at risk by institutions that wield even mere coordinative power. For instance, an institution setup to coordinate action around geoengineering could suggest that resources be directed toward a particular (perhaps comparatively less optimific) technology or could have a bias in favor of development. In such a situation, even if the institution didn't back its suggestions with coercive power, even its mere coordinative power could put us at risk of being locked into the development of a suboptimal technology or could even lead us toward objectionable deployment scenarios.

17. Buchanan, 193.
18. Buchanan, *The Heart of Human Rights*.
19. Buchanan, *The Heart of Human Rights*, 180.
20. The normative criterion of comparative benefit is, in fact, a necessary condition, and not merely a counting principle. Why is it the case that comparative benefit is a necessary condition and not merely a point in favor of an institution being considered legitimate? The answer is conceptual. Insofar as we have institutions in order to help solve coordination problems and to do so without the costs and inefficiencies associated with a non-institutional alternative, an institution that fails to deliver a comparative benefit is not helping to solve the original coordination problem. Now, it may be the case that institutions that go significantly beyond the minimal satisfaction of the criterion of comparative benefit ought to be considered more legitimate than institutions that barely satisfy such a condition. But when assessing an institution that fails to meet the criterion of mere comparative benefit, we need not even examine the other normative criteria. Any institution that fails the test of comparative benefit ought to be judged illegitimate.
21. There are good reasons not to be overly demanding and require perfect satisfaction of all the relevant normative criteria. For instance, sometimes institutions are required in order to further the cause of justice, and it would, ironically, undermine the cause of justice to require that all institutions be perfectly just in order for them to be considered legitimate.
22. See note 20 above.
23. Previously, I understood the comparative benefit criterion in a "thicker" sense than it is formulated here. Previously, I understood the comparative benefit criterion as meaning that an institution had to make the world a better place than it would be absent the institution in question. But this "thick" understanding led to a number of difficulties. I have since revised this view and now spell out the comparative benefit criterion in a much "thinner" way. The comparative benefit criterion now means only that the institution must enable coordination around institutional goals, and leaves the decision of what those goals actually are up the international community. I thank an anonymous reviewer for pushing me on this point.
24. Buchanan and Keohane, "The Legitimacy of Global Governance Institutions," 422.
25. Buchanan and Keohane, 426.
26. Ngaire Woods, "Holding Intergovernmental Institutions to Account," *Ethics and International Affairs* 17, no. 1 (2003): 69–80.
27. Maskin and Tirole, "The Politician and the Judge."
28. Take the case of genetically modified mosquitoes, for instance. Imagine that politicians, speaking with informed scientists, recognize that genetic modification is the all-things-considered best way to stop the spread of infectious disease. If there is too much accountability, they may be reluctant to pursue such a policy option because their constituency may be averse to genetic modification. Adrienne LaFrance writes, "A poll conducted by the Annenberg Public Policy Center in February found more than one-third of Americans believed genetically modified mosquitoes were to blame for the spread of Zika. (They're not)." See Adrienne LaFrance, "Genetically Modified Mosquitoes: What Could Possibly Go Wrong?," *The Atlantic*, April 26, 2016, https://www.theatlantic.com/technology/archive/2016/04/genetically-modified-mosquitoes-zika/479793/.
29. Gardiner, *A Perfect Moral Storm*, Chapter 5.
30. We will want to avoid "narrow accountability" in favor of a more *broad accountability*. Narrow accountability is "accountability without provision for contestation of the terms of accountability." Broad accountability, however, would allow for public discourse and require a fifth element for the criterion, namely, that the terms of accountability themselves be publicly

justified and open to revision. See Allen Buchanan and Robert O. Keohane, "The Legitimacy of Global Governance Institutions," *Ethics & International Affairs* 20, no. 4 (2006): 405–37.

31. For more on how to identify the relevant constituency for the Metacoordination View, see Pietro Maffettone and Luke Ulaş, "Legitimacy, Metacoordination, and Context-Dependence," *International Theory* 11 (2019): 81–109.

32. Etzioni, "Is Transparency the Best Disinfectant?"

33. Bannister and Connolly, "The Trouble with Transparency."

34. Buchanan and Keohane, "The Legitimacy of Global Governance Institutions," 427.

35. O'Neill, "Transparency and the Ethics of Communication."

36. Kevin Elliott argues that the scientific community has an obligation to disseminate information "in a manner that promotes the ability of those affected by the decisions to provide some form of informed consent." The normative criterion of transparency would work much the same way, allowing the public to either continue to grant the geoengineering governance institution the standing it needs to function, or deny it that standing. See Elliott, "An Ethics of Expertise," 642.

37. Not only can institutions actually deliver certain benefits and burdens, they can also affect the possibility of certain benefits and burdens coming to fruition. Whether an institution has actually produced particular benefits or burdens can often only be assessed ex post. For instance, if we assume that a geoengineering regulatory institution would deliver a comparative benefit, we should grant it standing ex ante. But if it fails to deliver such benefits and is actually putting us at risk of catastrophe, then we should collectively revoke its standing. We could also recognize that an institution could put us at risk ex ante, in which case we should also withhold a positive legitimacy assessment. See section 4.1 for more on risk and comparative benefit.

38. Buchanan, "Institutional Legitimacy," 55.

39. While it is not completely free from controversy, the following claim enjoys rather broad support. See Stephen M. Gardiner, "Ethics and Global Climate Change," *Ethics* 114, no. 3 (April 2004): 555–600; Stephen M. Gardiner, "Geoengineering: Ethical Questions for Deliberate Climate Manipulators," in *The Oxford Handbook of Environmental Ethics*, ed. Stephen M. Gardiner and Allen Thompson (Oxford: Oxford University Press, 2016); Joshua B. Horton and David Keith, "Solar Geoengineering and Obligations to the Global Poor," in *Climate Justice and Geoengineering: Ethics and Policy in the Atmospheric Anthropocene*, ed. Christopher J. Preston (London: Rowman & Littlefield International, Ltd., 2016), 79–92; and David R. Morrow, "Fairness in Allocating the Global Emissions Budget," *Environmental Values* 26, no. 6 (December 1, 2017): 669–91.

40. There are two ways to understand the claim that the least well off should be favored most heavily: in an absolute sense and in a relative sense. I think that substantive justice would require any use of climate engineering to favor the least well off in both a relative and an absolute sense. This entails that if the use of climate engineering were to make everyone worse off, but were to make the least well off less worse off than others, that it would not further the cause of justice. That is, I am not entertaining ideas of levelling down. I thank Stephen Gardiner and Augustin Fragniere for bringing this to my attention.

41. For more on how environmental burdens are often unjustly thrust upon vulnerable communities, see Cutter, *Hazards, Vulnerability, and Environmental Justice*; Bullard, *The Quest for Environmental Justice*; Taylor, D., *Toxic Communities*.

42. Rawls, *Political Liberalism*, 424; Hampshire, "Liberalism: The New Twist."

43. While he doesn't use these terms and focuses more on what I call fair terms of inclusion, these terms are influenced by Charles Beitz's discussion of political equality. See Charles R. Beitz, *Political Equality: An Essay in Democratic Theory* (Princeton, NJ: Princeton University Press, 1989).

44. For similar thoughts on what has been coined "participative justice," see Shrader-Frechette, *Environmental Justice*; Schlosberg, *Defining Environmental Justice*.

45. For the canonical discussion on descriptive or sociological legitimacy, see Max Weber, *The Theory of Social and Economic Organization* (New York: Free Press, 1997). For a comparison of descriptive and normative legitimacy, see my article on which this chapter is based: Daniel Edward Callies, "Institutional Legitimacy and Geoengineering Governance."

Chapter Five

Substantive Justice

§1 INTRODUCTION

Geoengineering has the potential to reduce the residual climatic harms associated with any realistic amount of mitigation and adaptation.[1] Thus, we can see geoengineering as potentially delivering a kind of benefit (along with, presumably, some burdens as well) compared to some baseline of fixed mitigation and adaptation efforts. To the extent that we want our governance institution to be guided by the normative criterion of substantive justice, we have to determine what a proper distribution of these benefits and burdens would look like. Assuming that everyone has an interest in securing the benefits and avoiding the burdens associated with geoengineering, what would be an equitable distribution of these benefits and burdens?

In his 1999 paper, Henry Shue argues that three common-sense considerations of fairness all point to the same conclusion about who should foot the bill for climate change mitigation.[2] In this chapter, I want to follow Henry Shue and argue that three (similar, yet different) considerations of fairness all point to the same conclusion regarding solar radiation management (SRM): if SRM is to be deployed, equity in our overall climate policy requires that the distribution of benefits and burdens of deployment must be skewed heavily in favor of those who are least well off. That is, substantive justice demands that the least well-off members of our global community should be the primary beneficiaries of any intentional intervention in the climate system.

The least well-off members of our global community ought to receive a greater share of the benefits and a smaller share of the burdens of SRM for at least the following three reasons: (1) they aren't causally responsible for the need to deploy geoengineering, that is, they aren't causally responsible for climate change, and thus have a strong claim against its associated burdens;

(2) they are not the primary beneficiaries of the actions that have brought about climate change, and so again have a strong claim against its burdens; and (3) they have the weakest ability to respond to the burdens of climate change, which gives them stronger claims to the benefits of geoengineering. These are relevant considerations of fairness for determining a proper balance between the competing claims to the benefits of geoengineering and they count toward skewing the balance in favor of those who are least well off.

The chapter will proceed as follows. The following section shows that we have many decisions to make with respect to SRM, and that these decisions will impact the distribution of benefits and burdens associated with deployment. Section 3 will look at why considerations of fairness are relevant to climate change policy. Sections 4–6 then each explore different considerations that speak in favor of the benefits of SRM being aimed at those least well-off members of the global community, and section 7 concludes the chapter.

§2 POTENTIAL BENEFITS (AND BURDENS) OF SRM

Deciding to engineer the climate with SRM is not a binary decision, despite it often being characterized as one. An analogy here will help. Imagine you have a room you wish to illuminate. The decision ahead of you is not a binary one of either having the room illuminated or not; you have decisions to make. Do you opt for overhead lighting or table lamps? If overhead lighting, should it be a hanging fixture or recessed? Do you want a warm, incandescent bulb or a colder fluorescent one? Note that there are a number of permutations available that will each create a novel lighting scenario for the room in question. Similarly, with SRM, decisions surrounding deployment are numerous and complicated. First, what kind of aerosol do we use: sulfuric acid, calcium carbonate, something else? Where do we inject the aerosols: at the equator, at 30°N, at 30°S? Do we engineer with the intention of offsetting the disruption to the global hydrological cycle or the disruption to average global surface temperature? If we go for temperature, do we engineer enough to offset the entire temperature anomaly, or just half, or some fraction thereof? These are just some of the many questions that will have to be answered if SRM is ever to be deployed at the global scale. And the decisions we make will affect the distribution of benefits and burdens produced by an engineered climate. The potential for differentiated impacts from SRM is well-documented, as the following examples illustrate.

Peter Irvine, Andy Ridgwell, and Daniel Lunt show how using SRM to offset any given percentage of the increase in average global surface temperature since the Industrial Revolution will have varied regional effects on precipitation anomalies. For example, under a scenario with no geoengineering and a quadrupling of the preindustrial level of atmospheric CO_2, the United States is expected to receive, on average, 7.7 percent more precipitation per year. Under the same scenario, the eastern region of China is expected to receive, on average, a 28.6 percent increase in precipitation; Australia is projected to have a 28 percent *decrease* in precipitation, and Brazil is expected to have 34.8 percent *decrease* in average yearly precipitation.[3] If the United States was interested in offsetting the increase in average yearly precipitation expected to be brought on by climate change, it would advocate for a geoengineering scheme that would offset 40–50 percent of average global surface temperature.[4] If China or Australia were interested in offsetting their increase in average yearly precipitation expected to be brought on by climate change, they would advocate for more geoengineering, since under a scenario with a quadrupling of CO_2 and enough geoengineering to offset the entire associated temperature increase they would experience nearly no precipitation anomaly. Further still, if solely interested in returning to preindustrial average precipitation values, Brazil would prefer a geoengineering scheme that offset even more than 100 percent of the expected rise in average global surface temperature associated with the quadrupling of preindustrial levels of atmospheric CO_2.[5] Thus, even if all countries agreed on the target of offsetting changes in average yearly precipitation—a target that would, of course, be irresponsibly myopic—they would prefer radically different amounts of geoengineering.

And what specific level of average global surface temperature we aim at is not by any means the only possible point of contention. We can also expect significant differences in climate depending upon how we achieve such a reduction in average global surface temperature. Haywood et al. demonstrate that where we choose to deploy geoengineering could have drastic effects on specific climatological features. Analyzing the potential for drought in Africa's Sahel region, Haywood et al. conclude:

> deliberate geoengineering injections into the *Northern Hemisphere* [alone] will preferentially load the Northern Hemisphere stratosphere *causing Sahelian drought*, whereas stratospheric geoengineering in the *Southern Hemisphere* [alone] will cause a significant *increase* in Sahel vegetation productivity or a greening of the region.[6]

Their study shows that even the decision of where to inject a particular aerosol can have profound regional effects, creating novel distributions of benefits and burdens.

Finally, the kind of aerosol used to enhance the Earth's albedo can result in differentiated burdens and benefits. As was noted earlier, one of the risks of injecting sulfate particles in the upper atmosphere is the likely untoward side effect of hindering ozone recovery. Decreased atmospheric ozone allows more UVB solar radiation to reach Earth's surface, leading to an increase in non-melanoma skin cancers.[7] Of all the countries on Earth, Australia suffers from the highest rate of nonmelanoma skin cancers,[8] which provides Australian officials with a reason to be warry of hindering atmospheric ozone recovery. They very well could be more interested in using calcium carbonate aerosols instead of sulfates, since calcium carbonate can actually *aid* ozone recovery rather than hinder it.[9] With these examples in mind, it is clear that we have choices to make with respect to SRM.[10] Even if we all agreed that deployment was a good idea—a scenario that is unlikely to ever come about—there would still be considerable disagreement regarding *how* SRM should be deployed.

§3 FAIRNESS IN INTERNATIONAL CLIMATE POLICY

When asking how SRM should be deployed, we can approach the question from different vantage points. For instance, we might ask, from the view point of economics, what deployment scenario would go furthest in minimizing the expected loss climate change is projected to have on global GDP. Or we might ask, from the perspective of ecology, what deployment scenario does the best job of preserving biodiversity. This chapter, however, centers on a different question. In attempting to elucidate the normative criterion of substantive justice, we ask: What would be the fairest or most equitable deployment of SRM? In other words, what would a distributively just deployment of SRM look like?

As has been mentioned, distributive justice can be thought of as a proper balance between competing claims to the benefits and burdens of social cooperation.[11] But one might immediately question whether considerations of justice are apt when focusing on anthropogenic climate change.[12] Do climate change and climate change policy really count as the kind of "social cooperation" needed to trigger considerations of distributive justice? Isn't the realm of international relations a war of all against all? Most theorists tend to think there are certain conditions that must be met in order for considerations of distributive justice to arise.[13] Philosophers such as Richard Miller,[14] Michael Blake,[15] and Thomas Nagel,[16] for instance, all claim that there must be relations of coercion—that is, that individuals must be under the same coercive state institutions—in order for considerations of distributive justice to arise. Others, such as Andrea Sangiovanni, suggest that when individuals stand in a relation of reciprocity to one another—that is, when they reciprocally maintain institutions that are necessary for a decent human life—then and only then are considerations of distributive justice appropriate. Darrel Moellendorf, among others, maintains that duties of distributive justice are triggered not by coercion or reciprocity, but rather by relations of a specific association—only when people are "co-participants in an association of the requisite kind"[17] are considerations of distributive justice appropriate.

If it could be shown that the right kind of triggers are met in the case of climate policy, then we could ground our evaluations of different distributions of the benefits and burdens of SRM in a specific conception of justice. For instance, if a legally binding climate treaty with significant coercive mechanisms were to be in effect, then it could be plausible to say that coercion of the kind needed to trigger considerations of justice for theorists such as Blake, Nagel, or Miller had been met, and we could then rely upon one of their theories of justice to guide policy. Likewise, if we were to see the participants of the UNFCCC as engaging in reciprocal action to maintain an institution that is "necessary for developing and acting on a plan of life,"[18] then we could embed our analysis of SRM in a reciprocity-based conception of justice, like the one advanced by Andrea Sangiovanni. Or if the association of 197 states aiming to protect the climate system is the right kind of association that triggers considerations of justice for the kind of conception advanced by Darrel Moellendorf, then we could simply analyze SRM through that lens and arrive at a conclusion about the fair distribution of the technology's benefits and burdens.

If we were to take any of these routes, we would encounter a hurdle. Relying upon a particular conception of justice to inform our thoughts on climate policy requires what Moellendorf calls a *deep justification*, or a justification "based on a comprehensive account of justice that applies to all of the most important aspects of global justice."[19] Relying on such a deep justification has at least two drawbacks. The first is such reliance on a specific conception of global justice would require thorough and significant justification—justification that would require a book all to itself, making it beyond the scope of this work. Second, the defense of any particular conception of justice will surely fail to convince everyone. This would mean that the evaluation of any deployment scenario "would stand or fall with the particular theory of justice on which it is based."[20]

Fortunately, we need not rely upon a comprehensive conception of global justice to ground the idea that fairness is important when evaluating the distribution of benefits and burdens of potential SRM deployment scenarios. Following Moellendorf,[21] I am going to ground the reliance upon considerations of fairness in the relatively noncontroversial idea that climate policy ought to be governed by the norms already agreed to in the United Nations Framework Convention on Climate Change. In 1992, 197 countries all committed to "protect the climate system for the benefit of present and future generations of humankind, on the basis of *equity* and in accordance with their *common but differentiated responsibilities and respective capabilities*."[22] As Henry Shue points out, "What diplomats and lawyers call equity incorporates important aspects of what ordinary people everywhere call *fairness*."[23] Assuming that SRM would be part of our overall climate policy, and assuming that we recognize the norms embedded in the UNFCCC constrain our climate policy with considerations of fairness, the question is: What would be a fair distribution of the benefits and burdens of SRM? The following sections will outline three considerations of fairness that count in favor of making sure that the burdens of SRM are not thrust upon those who are least well off.

§4 CAUSAL RESPONSIBILITY

If climate change were a natural phenomenon, meaning that no human or group of humans could justifiably be attributed responsibility for its genesis, then our evaluation of geoengineering and climate policy in general would be starkly different. The climate change we are experiencing and will continue to experience in the future, however, is not a natural phenomenon. There is, of course, natural variability in the climate, but, that natural variability notwithstanding, the main driver behind the increase in average global surface temperature in the past century is clear. The climate change we are experiencing (and will continue to experience into the future) is driven by anthropogenic greenhouse gases. "Their effects, together with those of other anthropogenic drivers, have been detected throughout the climate system and are *extremely likely* to have been the dominant cause of the observed warming since the mid-20th century."[24]

If it is anthropogenic greenhouse gases that are almost entirely responsible for climate change, then who is responsible for the accumulation of greenhouse gases in the atmosphere? When talking about responsibility for climate change, it is important to keep certain distinctions in mind. The first pertinent distinction is that between moral and causal responsibility. When I say, "The drought was responsible for the crop failure" or "The waves were responsible for eroding the coastline," what I mean to say is there was a relation of cause and effect. I mean to say that the drought *caused* the crops to fail and that the waves *caused* the erosion of the coastline. These are purely descriptive statements that do not necessarily carry any moral weight. After all, it would be absurd to assign moral praise or blame to things like droughts or waves. And while statements of causal responsibility can be appropriate for inanimate objects and events, statements of causal responsibility can also be made about moral agents. I can say, "Dustin fired the gun," which means simply that Dustin caused the gun to fire. As just mentioned, statements of causal responsibility do not necessarily assign praise or blame, even when uttered about moral agents.

Who is causally responsible for the buildup of greenhouse gases in the atmosphere? The answer that is commonly given is: the early-industrialized, wealthy countries of the world. Henry Shue writes, "The industrial activities and accompanying lifestyles of the [developed countries] have inflicted major global damage upon the earth's atmosphere."[25] Stephen Gardiner claims that "the responsibility for historical and current emissions lies predominantly with the richer, more powerful nations."[26] In the same vein, Eric Neumayer says "global warming is caused by cumulative emissions and the developed countries have contributed much more to cumulated emissions than the developing world."[27] In fact, the idea that the developed nations of the world are primarily causally responsible for the increased concentration of greenhouse gases in the atmosphere is almost universally recognized. The third paragraph in the preamble to the United Nations Framework Convention on Climate Change reads, "The largest share of historical and current global emissions of greenhouse gases has originated in developed countries."[28]

If we look at causal historical responsibility for the accumulation of greenhouse gases in the atmosphere, here is what we find. Using data derived from the World Resources Institute's Climate Analysis Indicator Tool (CAIT), we can see that roughly 750,000 Mmt CO_2 have been released in between the years of 1850 and 1990.[29] Now, while a complete disaggregation of these emissions is impossible, a relatively accurate disaggregation has been calculated and divided up among various nations. Using the United States, the European Union, China, India, Brazil, and Sierra Leone as examples, the data is quite explicit about who is causally responsibly for the rise in atmospheric CO_2 concentrations over the past century and a half. While the United States is currently one of the highest overall and per capita emitters, the states of Europe have a longer history of industrialization and thus have almost identical emissions over the 150-year period. Now, there have obviously been many other players involved over the past years, but both the United States and the European Union are each responsible for about 32 percent of the historical emissions up to 1990. China weighs in with only 5 percent, India with 1.5 percent and Brazil reaches a mere 0.6 percent of emissions prior to 1990.[30] Most importantly, even though Sierra Leone has contributed something over the past years, its contribution is relatively so miniscule that it does not earn it one-tenth of a percent of causal responsibility.

If we were to grant this causal responsibility moral relevance when assessing the claims to the potential benefits of SRM, the conclusion would be rather clear. It would be unfair to deploy geoengineering under a scenario in which the benefits of the technology went to those in the industrialized countries who caused the problem, and even more unfair to pass further burdens along to those in the developing countries of the world who bear little causal responsibility for the buildup of greenhouse gases in the atmosphere.[31]

There are two different ways in which moral relevance or moral responsibility could be connected to causal responsibility. The moral relevance of causal responsibility can be divided into two sub-cases: fault and no-fault.[32] To illustrate how causal responsibility can generate no-fault moral responsibility, imagine you park your car on the street overnight. Suppose your neighbor, exercising due caution and genuinely attempting to safely guide her car down the road, happens to swipe your side-view mirror clean off your car. There is no doubting her causal responsibility for the damage done to your car, but she didn't intend such damage and was in no way acting negligently. Accidents, unfortunately, happen. Despite her driving with appropriate care and not even remotely intending to damage your car, we nonetheless think she is morally responsible for the damage she has caused. This is an example of no-fault moral responsibility.[33] She is not blameworthy for her actions, but they are nonetheless morally relevant when assessing what should be done about the broken side view mirror.

Contrast this example of no-fault moral responsibility with the following. You park your car on the street overnight. Having just enjoyed five delicious Sculpin Indian Pale Ales at the Ballast Point Brewery, I decide I am going to, despite my drunkenness, drive home. I manage to make it all the way to our neighborhood without an accident. But, just when I am almost home, I fall asleep at the wheel and swipe your side view mirror clean off your car. Just as in the first case, I am undoubtedly causally responsible for the damage done to your car. But in this case, not only am I causally responsible for the damage, I am also morally responsible and at fault. That is to say that in addition to being morally responsible for the cost of replacing the side-view mirror, I am also blameworthy for my actions.

These examples highlight some of the important conditions for attaching fault to an agent casually responsible for an outcome. Moellendorf argues there are four relevant conditions for attributing fault to a party:

> (1) outcome—the outcome must be credited to the agent; (2) care—the agent's action must have contravened a standard of care, to which it is reasonable to hold an agent; (3) voluntariness—the agent must have acted voluntarily; and (4) knowledge—the agent must have acted with knowledge about the likely outcomes of her action.[34]

Now, at least two problems with attributing fault to those in the developed countries of the world for their excessive emissions become immediately evident. First, the fourth condition mentioned above by Moellendorf—that of knowledge of the likely outcomes—does not hold. An agent must have knowledge of the harm she is causing in order to be assigned fault for an action. Knowledge of climate change and its genesis was anything but widespread prior to 1990. This being the case, it seems doubtful that we can hold people or nations morally at fault for a harmful action to which they had no knowledge they were contributing. Second, we can't say that those in the industrialized countries were failing to show an appropriate standard of care either. Without knowledge of the harmful effects of an overabundance of greenhouse gases in the atmosphere, it is unreasonable to say that any duty of care was violated. For at least these two reasons, it would seem that the basis for holding the wealthy nations at fault for their historical emissions is undermined.

But what of a no-fault account of the moral relevance of past emissions? A no-fault account of past emissions would be somewhat similar to the old adage of "you break it, you buy it." It's not that you are at fault or morally blameworthy for breaking, say, an expensive vase in a flower shop. After all, accidents happen. But we do think your causal connection to the broken vase is morally relevant when deciding who should bear the costs of replacement. This is the idea behind strict liability. Under this no-fault account, the agent or agents causally responsible for an event are held *strictly liable* for its consequences and the costs those consequences may induce. This seems appropriate for the subject matter of climate change. After all, not only were people acting unintentionally when emitting GHGs into the atmosphere, but the fact that this action would bring about harm also depended upon circumstances outside of their control, such as the heat-trapping properties of the greenhouse gases they were emitting. Nonetheless, there are harms that are resulting from their actions, and it would be unfair to ignore this when negotiating our climate policy.[35]

So, under an account of strict liability, states are not at *fault* for their historical emissions—they are not blameworthy—but they are *morally responsible* for the harm these emissions will bring about. They are morally responsible in the sense that the driver in our previous example is responsible for the damage she caused to her neighbor's car. Holding states strictly liable for their historical emissions places moral weight on the fact that those in wealthy countries are the ones who have caused the problem, and thus their claim to have the burdens of climate change alleviated through SRM are weaker than the claims of those in developing countries. Fairness seems to require us to place at least some moral weight on this consideration.

However, there is one significant difficulty with holding the members of wealthy states liable for historical emissions. The vast majority of those causally responsible for historical emissions are now dead. It could be objected that it is unfair to hold current citizens of the early-industrialized world liable for actions taken before they were even born. But, even if those in wealthy countries reduced their emissions to zero today—a feat that has yet to happen and is not projected to occur in this half of the century[36]—there may still be a morally relevant connection between current citizens and these emissions prior to their comings into existence. This morally relevant connection is the topic of the next section.

However, even if we decide to ignore past emissions and look at more recent contributions to the stock of greenhouse gases in the atmosphere, we still find it is the citizens of the wealthy, industrialized states that are primarily responsible for the continued buildup of heat-trapping gases.[37] Looking at per capita emissions in 1990 when the IPCC released its first assessment report, we see that citizens in high-income countries were still releasing much more than the global mean. On average, citizens in high-income countries were releasing 11.5 metric tons of CO_2 in 1990. Compare that to the average emission of 2.1 metric tons of CO_2 from citizens in low- and middle-income countries.[38] And if we fast-forward to today, we see that citizens of wealthy countries are still emitting, on average, more than those in low- and middle-income countries. While the emissions of those in high-income countries have decreased since 1990, they are still well above the global average. The average yearly emissions of those in high-income countries was still 11 metric tons of CO_2 in 2013, more than triple that of the average of 3.5 metric tons of CO_2 from an individual in a low- or middle-income country.[39] Thus, even if we ignore emissions prior to 1990, we see that citizens in developed countries have been contributing disproportionately to the continuing buildup of greenhouse gases in the atmosphere. To the extent that we think such a continued disproportionate contribution to the problem has moral relevance when assessing the claims people have to the benefits of climate remediation technology, we should be confident in claiming that those in the developing world have stronger claims to the benefits that SRM can offer and that any deployment of the technology ought to favor them on grounds of fairness.

§5 BENEFICIARY RESPONSIBILITY

The previous section outlined the disproportionate historical contribution to climate change on the part of those residing in developed countries. It was suggested that we consider these historical emissions morally relevant when determining the proper balance between the competing claims of current individuals to the benefits of geoengineering. But it was noted that some may question the moral relevance of past emissions on grounds of fairness. The thought was that it would be unfair to hold current generations morally responsible for the emissions of their ancestors. Side-stepping this objection, it was shown that even a narrow focus on more current emissions after 1990 led to the same conclusion. Even current members of the developed world are disproportionately causally responsible for anthropogenic climate change compared to those in the developing world, and this causal responsibility is morally relevant.

But abandoning historical contributions to climate change and relying merely upon current causal responsibility when assessing the fairness of our overall climate policy faces a countervailing objection.[40] If we disregard emissions prior to 1990, then at some point in the future we will have to conclude that it is those in the developing world who are responsible for greater contributions to climate change. After all, a child born in Sweden today will release on average 4.6 metric tons of CO_2 each year, with that number *decreasing* as time passes. A child born in Equatorial Guinea today will release on average about 6.8 metric tons of CO_2 each year, with that number *increasing* as time passes.[41] Completely disregarding emissions prior to 1990 and focusing on greenhouse gases that have been emitted with knowledge of their harmful effects would have us conclude that those in the developing world, in their attempt to escape crippling poverty, will soon be the ones primarily responsible for climate change.[42]

For this reason, many think it would be morally unacceptable to ignore previous emissions. In an attempt to hang on to historical emissions, some ground their moral relevance in the idea of *beneficiary* responsibility. Like the idea of strict liability, beneficiary responsibility is also a no-fault account. However, unlike strict liability, beneficiary responsibility does not require any causal connection to the genesis of the problem. Beneficiaries of the actions that have brought about climate change are not causally responsible, but they nonetheless have a connection to the cause of the problem through the relation of benefiting from it. The simple thought associated with beneficiary responsibility, then, is that those who have benefited from the actions that cause climate change are morally responsible for the burdens it is going to bring about. This responsibility would, in turn, make their claims to relief from climatic burdens through SRM weaker than the claims of those who have not enjoyed such benefits.

Who is it that has benefited from the historical emissions that are largely responsible for climate change? Henry Shue answers that question with another question. He asks: What is the difference between being born today in Belgium and being born today in Bangladesh?

> Clearly one of the most fundamental differences is that the Belgian infant is born into an industrial society and the Bangladeshi infant is not. Even the medical setting for the birth itself, not to mention the level of prenatal care available to the expectant mother, is almost certainly vastly more favourable for the Belgian than the Bangladeshi. Childhood nutrition, educational opportunities and life-long standards of living are likely to differ enormously because of the difference between an industrialized and a non-industrialized economy. In such respects current generations are, and future generations probably will be, continuing beneficiaries of earlier industrial activity.[43]

This attribution of benefits is echoed by Peter Singer who writes, "the [current] wealth of the developed nations is inextricably tied to their prodigious use of carbon fuels."[44] And Eric Neumayer writes, "There can be no doubt that the development of the 'Northern' countries was eased, if not made feasible in the first place, by having had the possibility of burning large amounts of fossil fuel with the consequence of an accumulation of carbon dioxide in the atmosphere."[45]

Now, the conclusion that the benefits of historical emissions have accrued to the members of the wealthy, early-industrialized states can be contested. One might agree that *some* of the benefits have gone to those in the developed countries, but that those residing in developing countries have surely benefited as well. As Neumayer points out, Grubb et al. "argue that past emissions enabled the development of public goods such as modern medicine or better technologies that have also raised living standards in developing countries and make it easier for later developing countries to gain the same living standards with less emissions."[46]

The thought that individuals outside of high-emitting states have also benefited from those historically high emissions is uncontestable. But so, too, is the thought that, while *some* benefits have leaked out to members of other countries, the vast majority of the benefits of historical emissions are enjoyed by individuals born into already developed countries. If we conceive of the benefits of such historical emissions as monetary income, it's clear that those living in countries with historically high per capita emissions now enjoy significantly higher incomes than those living in countries with historically low per capita emissions. Janssen et al., for example, demonstrate that countries' past contributions to atmospheric CO_2 concentrations can account for up to two-thirds of the current variations in per capita GDP.[47] And if we were to use human development as an index for calculating the benefits of historical emissions, we see that no country in the group of those labeled as "very highly developed" has been able to get there without significant per capita emissions in the past. South Korea and Swaziland are perfect examples. In the thirty-year period from 1975 to 2005, South Korean per capita emissions rose from 2.3 metric tons of CO_2 to 9.7 metric tons; contemporaneously, the country's Human Development Index (HDI) score rose from .713 to .921. Swaziland, on the other hand, has not been so fortunate. In the thirty-year period from 1975 to 2005, Swaziland's per capita emissions barely rose from 0.63 metric tons of CO_2 to 0.9 metric tons; contemporaneously, the country's HDI score rose merely from .527 to .547. The relation of per capita emissions to HDI is certainly far from perfect—a number of factors other than per capita emissions contribute to a country's development. But it is also clear that with its higher per capita emissions, South Korea was able to reach a high level of development, while Swaziland was not. Different studies have concluded that, while not sufficient, high per capita emissions are a necessary condition for human development—at least until cleaner energy becomes more readily available and cheaper.[48] So while the benefits of GHG emissions may leak out to other corners of the globe, evidence suggests that the relation between historical emitters and current beneficiaries is a strong one.

Relying upon beneficiary responsibility to reach the conclusion that the benefits of SRM should be directed toward those in the developing world faces another challenge. Some think that holding beneficiaries responsible for the actions of others, even if they have benefited from such actions, is insufficient to ground any moral responsibility on the side of the beneficiary. For instance, imagine my neighbor fixes up her house, thereby raising the property value of my home too. I have not participated in the action in any way, and yet I am undeniably a beneficiary; I am enjoying a positive externality of her work. But my benefitting from her home improvement does not seem morally relevant. When I go to sell my home, I in no way have a moral responsibility to share some of the extra money I'll receive with her or anyone else. Merely being a beneficiary of someone else's action is insufficient to ground moral responsibility; there must be more to the picture.[49]

Most theorists think that, for beneficiary responsibility to be morally salient, the action from which one benefits must have been either wrongful or unjust. We have already established that wealthy states should not be held at fault or blamed for their early development that relied upon intensive greenhouse gas emissions. So it does not seem as if present-day members of wealthy states are the beneficiaries of any *blameworthy* consumption of the global atmospheric sink. But the disproportionate consumption of the atmosphere's absorptive capacity by the early-industrialized states does seem to constitute a kind of *unjust enrichment*, to use the language of Edward Page.[50] There has been an imbalance in the distribution of the benefits of a common global commodity. Page explains:

> The generation of significant benefits, which became concentrated in the developed states, both in the form of accumulated wealth and national income, can be traced to the exploitation of the storage and sink capacity of the climate system, which itself should be viewed as commonly, or jointly, owned by all states. . . . The behavior of developed states since 1750 can be conceived as an instance of accidental, but nevertheless profitable, trespass on the atmospheric commons the value of which should be spread across all states more evenly than is presently the case.[51]

Previous generations' reaping of the benefits of industrialization and greenhouse gas emission while passing the burdens onto the poor of current and future generations—notwithstanding their ignorance of the situation—should be considered an injustice given that it is making it difficult for current and future generations to even meet their basic needs.[52] The fact that those in the developed world have (unjustly) benefited more from the historical emissions that are still to this day driving climate change is morally relevant when assessing claims to the benefits and burdens of the deployment of geoengineering. Their beneficiary responsibility weakens their claims to the benefits of SRM. Conversely, the fact that those in the developing world have benefited the least from the activities that are bringing about climate change provides us with a reason to count their claims to the benefits of SRM as comparatively stronger.[53]

§6 ABILITY

Notions of both causal and beneficiary responsibility push us toward the conclusion that SRM ought to be used to benefit the least well-off members of the globe—generally speaking, those in the developing world. But putting aside the developed countries' disproportionate casual responsibility for climate change, and putting aside the fact that they have (unjustly) benefited from the historical emissions of those now long dead, we have another reason to weight their claims to the benefits of SRM less heavily. One of those reasons rests on another no-fault conception of responsibility: the idea of differentiated abilities to respond to climate change.

We know that those in developing countries are going to be hit harder by climate change. But imagine that every country were to experience the same negative effects of a warmed world. And imagine again that climate change were not attributable to human activities, but were an unfortunate natural phenomenon. Even if that were so, a commitment to equity in our climate policy would require us to consider the developing countries' lack of ability to respond to the problem as a relevant factor. If we were motivated by a shortsighted conception of equality, we might assign the benefits and burdens of SRM as equally as possible across different populations. But a commitment to *equity* and the recognition of differentiated capacities to cope with the changing climate gives rise to claims of differing strengths. Those in developing countries would have a greater claim to the benefits of SRM because they have less of an ability to respond to climate change. That is, even if the climatic effects that were to visit the globe were equally distributed, someone in a developing country taking on the same climatic effect as someone in a developed country would translate into a much heavier burden, contravening the idea of fairness or equity.

Again, the idea of differentiated abilities does not attribute fault to those with greater ability to respond to the problem. Rather, it attributes moral relevance to their greater ability in the same way that a country with a progressive tax system attributes moral relevance to the wealth or income of its citizens when deciding upon the distribution of benefits and burdens of social programs. The wealthier members of a country with a progressive tax system are taxed more not because they are at fault for the relative lack of wealth of others within their country. Rather, they are taxed more either because it is thought they have benefited more from the social cooperation made possible by the state, or perhaps simply because they have the *ability* to contribute more to the provision of important social programs. Under either explanation, they are not *at fault* for the relative lack of wealth and lower incomes of their compatriots, but they are held *morally responsible* for their (larger) share of taxes. To wit, we think their greater wealth is morally relevant when dividing up the national tax burden.

To the extent that we recognize climate policy as a cooperative venture, and to the extent that we want this cooperative venture to be regulated by norms of equity, we should take the varying abilities of states into account when assessing how the benefits and burdens of SRM ought to be distributed. But even if we accept the idea that a country's or an individual's ability to cope with climate change is morally salient when assessing the proper distribution of benefits and burdens of SRM, an ancillary question immediately arises. How should we measure a country's or an individual's ability or lack of ability to respond to climate change? One straightforward suggestion would be to look at wealth. Climate change will carry with it many financial burdens, and the more disposable wealth one has available, the more one will be able to cope with those financial burdens. If we were to use wealth as an indicator of ability to respond to climate change, we could look at GDP per capita to determine which countries had the greatest ability to respond to climate change and which countries lacked such ability. We would find that countries like Luxembourg, Switzerland, and Norway have the greatest ability to respond while countries like South Sudan, Burundi, and Malawi are least able.[54] This would imply that Luxembourg, Switzerland, and Norway—along with other countries with high GDP per capita—would have the weakest claims to the potential benefits of SRM, while South Sudan, Burundi, and Malawi—along with other countries with meagre per capita GDP—would have the strongest claims to the benefits of geoengineering.

Despite being a reasonable place to start, wealth is not the best indicator of one's ability to cope to climate change. An even better indicator of ability to cope with climate change would be the idea of human development. Given that it takes into account more aspects of what we think broadly constitute the idea of ability to shoulder burdens and forego benefits, the Human Development Index (HDI) is a better way to measure the strength of claims to benefits and burdens of SRM.[55] If we were to rely upon the HDI, we would determine that countries with very high human development have weaker claims to the benefits of SRM than countries with low human development. This would imply that countries like Norway, Australia, and Switzerland— the three countries with the highest HDI ranking for 2015—have weaker claims to the benefits of SRM than countries like Niger, the Central Africa

Republic, and Eritrea—the three countries with the lowest HDI ranking for 2015.

The HDI is certainly a better indicator of ability to respond to climate change than a narrow focus on wealth. But there are at least two immediate problems with relying exclusively upon either wealth or human development to determine those with the strongest claims to the benefits of SRM. The first is that both wealth and development fail to take into account that individuals in different countries will have varying capacities to translate their wealth or development into an ability to respond to climate change. There would be have to something akin to the idea of purchasing power parity with respect to the conversion of these metrics to an ability to deal with climate change. The second and more pressing objection to using wealth or human development as an indicator is that it completely disregards exposure to harmful climatic events. For instance, according the International Monetary Fund, both the Netherlands and Austria boast a GDP per capita of roughly $45,000 for 2016 and both are in the HDI's group of countries with "very high human development."[56] However, while both countries have nearly identical numbers for per capita GDP and similar human development, the Netherlands will face significantly different exposure to the effects of climate change through sea-level rise, something a land-locked country like Austria does not have to worry as much about.

Perhaps the best measure of the strength of one's claims to the benefits and burdens of SRM is a measure of one's *vulnerability* to climate change. The more vulnerable one is to climate change, the stronger one's claims to the benefits of SRM; conversely, the less vulnerable one is, the weaker one's claims to the benefits of SRM. Researchers at the University of Notre Dame have compiled what they call the Global Adaptation Index that measures, among other things, varying vulnerabilities. According to the index, vulnerability "measures a country's exposure, sensitivity and ability to adapt to the negative impact of climate change. ND-GAIN measures the overall vulnerability by considering vulnerability in six life-supporting sectors—food, water, health, ecosystem service, human habitat and infrastructure."[57] According to the Notre Dame Index, those countries most vulnerable to climate change are Chad, Burundi, Somalia, the Central Africa Republic, and Eritrea; the countries least vulnerable to climate change are the United Kingdom, Germany, Denmark, Norway, the United States.[58]

If we were to attribute moral salience to the idea of vulnerability, we would conclude that those residing in the countries that are most vulnerable to climate change have the greatest claim to the potential benefits of SRM. There is no fault being attributed to those residing in the least vulnerable countries. Rather, we are simply maintaining that, as a matter of fairness, those who are most vulnerable have greater claims to the potential benefits and greater claims against the potential burdens of SRM. Fairness would thus require us to choose a distributive scenario in which the benefits are directed primarily toward individuals residing within vulnerable countries and the burdens are directed away from them.

Now, I have not offered a convincing argument in favor of considering vulnerability as the proper interpretation of one's ability to respond to climate change. It seems intuitively plausible, and certainly seems a better metric than per capita income or HDI ranking. Nonetheless, I assume that there is a correct (or least reasonably acceptable) standard for measuring one's ability to respond to climate change. Whatever that proper standard is, it is relevant for determining the appropriate balance between the competing claims to the benefits and against the burdens associated with geoengineering. Those who have the least ability to respond to climate change ought to receive a greater share of the benefits of geoengineering, and ought to be sheltered from its associated burdens.

§7 CONCLUSION

In this chapter I have given some content to the normative criterion of substantive justice as it relates to geoengineering. I argue that three considerations of fairness all count in favor of distributing the benefits and burdens of SRM most heavily in favor of the least well-off members of the globe. I want to emphasize the importance of this convergence. There are few in the philosophical literature who argue that those in the developing world morally deserve to shoulder the greater burdens of climate change. There is significant disagreement about why that is the case. For instance, some argue that those in the developing world should be relieved of some of the burdens they will face due to their lack of historical responsibility for the problem. Some argue they have claims against climatic burdens due to the fact that they have benefited the least from the actions that are driving climate change. Others argue it is their lack of ability to shoulder such burdens that gives them stronger claims against them. It is important to note that, while identifying different grounds for their conclusions, nearly everyone within the philosophical literature arrives at the same conclusion: the excessive burdens climate change places on those in the developing world is unfair.

Given the unfairness of the burdens that are being thrust upon those in the developing world, this chapter argued that they have a stronger claim for those burdens to be alleviated through SRM. In other words, this is what substantive justice requires of geoengineering. It is possible that you find the argument from causal responsibility irrelevant to climate policy. Perhaps it's the idea of beneficiary responsibility that seems inapt. But when various considerations, all identifying different grounds, deliver us to the same conclusion, that conclusion should not be ignored. I am not sure of what to say about a counterfactual scenario in which considerations of casual responsibility identified one group, beneficiary responsibility identified a different group, and ability pointed toward yet another group as the morally appropriate beneficiaries of geoengineering. Fortunately for our analysis, all roads have led to Rome. Whether we place moral weight on the idea of causal responsibility, beneficiary responsibility, or ability, we reach the same conclusion. For geoengineering to conform to the normative criterion of substantive justice, the deployment scenario we endorse should significantly favor those who are least well off.

NOTES

1. Toby Svoboda et al., "The Potential for Climate Engineering with Stratospheric Sulfate Aerosol Injections to Reduce Climate Injustice," *Journal of Global Ethics*, forthcoming (n.d.).
2. Henry Shue, "Global Environment and International Inequality," *International Affairs* 75, no. 3 (July 1, 1999): 531–45, https://doi.org/10.1111/1468-2346.00092.
3. Peter J. Irvine, Andy Ridgwell, and Daniel J. Lunt, "Assessing the Regional Disparities in Geoengineering Impacts," *Geophysical Research Letters* 37, no. 18 (September 2010): 1–6, https://doi.org/10.1029/2010GL044447.
4. Irvine, Ridgwell, and Lunt, 4.
5. Irvine, Ridgwell, and Lunt, 4.
6. Jim M. Haywood et al., "Asymmetric Forcing from Stratospheric Aerosols Impacts Sahelian Rainfall," *Nature Climate Change* 3, no. 7 (March 31, 2013): 663, https://doi.org/10.1038/nclimate1857 (emphasis added).
7. Environmental Protection Agency, "Health and Environmental Effects of Ozone Layer Depletion," Reports and Assessments, July 17, 2015, https://www.epa.gov/ozone-layer-protection/health-and-environmental-effects-ozone-layer-depletion.
8. Alexander Lomas, J. Leonardi-Bee, and F. Bath-Hextall "A Systematic Review of Worldwide Incidence of Nonmelanoma Skin Cancer," *British Journal of Dermatology* 166, no. 5 (May 2012): 1069–80.
9. A. Lomas, J. Leonardi-Bee, and F. Bath-Hextall, "A Systematic Review of Worldwide Incidence of Nonmelanoma Skin Cancer," *British Journal of Dermatology* 166, no. 5 (May 1, 2012): 1069–80, https://doi.org/10.1111/j.1365-2133.2012.10830.x.
10. I should mention that all of these examples are based on computer modelling, and not on direct observations. It is entirely possible that the observed impacts of these different scenarios will be different than what our models predict. But it is unlikely that the observed impacts would be so different so as not to generate differential benefits and burdens.
11. John Rawls, *A Theory of Justice* (Cambridge, MA: Harvard University Press, 1999), 9.
12. For an argument as to why considerations of justice are ill-advised when assessing climate policy, see Eric A Posner and David Weisbach, *Climate Change Justice* (Princeton, NJ: Princeton University Press, 2015); For a critique of their position and arguments in favor of incorporating considerations of justice in climate policy, see Darrel Moellendorf, *The Moral Challenge of Dangerous Climate Change: Values, Poverty, and Policy* (New York: Cambridge University Press, 2014).
13. For notable exceptions of so-called practice-dependent conceptions of justice, see Simon Caney, *Justice beyond Borders: A Global Political Theory* (Oxford: Oxford University Press, 2006) and Merten Reglitz, "Fairness to Non-Participants: A Case for a Practice-Independent Egalitarian Baseline," *Critical Review of International Social and Political Philosophy* 20, no. 4 (July 4, 2017): 466–85.
14. Richard W. Miller, "Cosmopolitan Respect and Patriotic Concern," *Philosophy and Public Affairs* 27, no. 3 (July 1998): 202–24, https://doi.org/10.1111/j.1088-4963.1998.tb00068.x.

15. Michael Blake, "Distributive Justice, State Coercion, and Autonomy," *Philosophy & Public Affairs* 30, no. 3 (July 1, 2001): 257–96, https://doi.org/10.1111/j.1088-4963.2001.00257.x.

16. Thomas Nagel, "The Problem of Global Justice," *Philosophy & Public Affairs* 33, no. 2 (2005): 113–147.

17. Darrel Moellendorf, "Equal Respect and Global Egalitarianism," *Social Theory and Practice* 32, no. 4 (2006): 601.

18. Andrea Sangiovanni, "Global Justice, Reciprocity, and the State," *Philosophy & Public Affairs* 35, no. 1 (2007): 4.

19. Moellendorf, *The Moral Challenge of Dangerous Climate Change*, 140.

20. Toby Svoboda, *The Ethics of Climate Engineering: Solar Radiation Management and Non-Ideal Justice* (New York: Routledge, 2017), 37.

21. Moellendorf, *The Moral Challenge of Dangerous Climate Change*, 136–40.

22. United Nations, "United Nations Framework Convention on Climate Change," 1992, http://unfccc.int/essential_background/convention/items/6036.php (emphasis added).

23. Shue, "Global Environment and International Inequality," 531.

24. Intergovernmental Panel on Climate Change, *IPCC, 2014: Summary for Policymakers. In: Climate Change 2014: Impacts, Adaptation, and Vulnerability. Part A: Global and Sectoral Aspects. Contribution of Working Group II to the Fifth Assessment Report of the Intergovernmental Panel on Climate Change* (Cambridge: Cambridge University Press, 2014), 4, http://www.ipcc.ch/pdf/assessment-report/ar5/wg2/ar5_wgII_spm_en.pdf (original emphasis). And when the IPCC says something is "extremely likely," they mean to say there is greater than a 95 percent probability in the statement's accuracy.

25. Shue, "Global Environment and International Inequality," 534.

26. Stephen M. Gardiner, "A Perfect Moral Storm: Climate Change, Intergenerational Ethics and the Problem of Moral Corruption," *Environmental Values* 15, no. 3 (August 1, 2006): 402, https://doi.org/10.3197/096327106778226293.

27. Eric Neumayer, "In Defence of Historical Accountability for Greenhouse Gas Emissions," *Ecological Economics* 33, no. 2 (2000): 187.

28. United Nations, "United Nations Framework Convention on Climate Change," Preamble.

29. This number is a raw measure of CO_2 emissions. It does not take into account the absorptive capacity of the atmosphere or other sinks.

30. World Resources Institute, "CAIT Climate Data Explorer," accessed August 18, 2017, http://cait.wri.org/.

31. Of course, there are some residing in the highly-developed states of the world who have contributed little to the problem. And, conversely, there are some residing in developing states who have contributed significantly to the problem. These outliers make a perfect distribution, even once identified, difficult to instantiate. But I set this pragmatic concern aside for now.

32. For the classic discussion of causal responsibility, moral responsibility, outcome responsibility, and fault, see Tony Honoré, *Responsibility and Fault* (Oxford: Hart Publishing, 2002); See also David Miller, *National Responsibility and Global Justice*, Oxford Political Theory (Oxford: Oxford University Press, 2012).

33. Try to ignore the fact that we would say "she is at fault for the accident."

34. Moellendorf, *The Moral Challenge of Dangerous Climate Change*, 165.

35. Strict liability has precedence in the environmental realm as it has been variably endorsed by the OECD and the European Union. See Simon Caney, "Cosmopolitan Justice, Responsibility, and Global Climate Change," *Leiden Journal of International Law* 18, no. 4 (January 9, 2006): 747. This is because, in many cases, it may be too difficult to prove a polluter is at fault, and thus strict liability is seen as a policy that is better able to provide the victims of pollution with the compensation they deserve. See David Weisbach, "Negligence, Strict Liability, and Responsibility for Climate Change | Belfer Center for Science and International Affairs," *The Harvard Project on International Climate Agreements*, 2010.

36. World Resources Institute, "Per Capita CO2 Emissions For Select Major Emitters, 2007 and 2030 (Projected) | World Resources Institute," accessed August 18, 2017, http://

www.wri.org/resources/charts-graphs/capita-co2-emissions-select-major-emitters-2007-and-2030-projected.

37. This is especially true if we discriminate between what Henry Shue calls "luxury emissions" and "subsistence emissions." See Henry Shue, "Subsistence and Luxury Emissions," *Law and Policy* 15, no. 1 (1993): 39–59.

38. World Bank, "CO_2 Emissions (Metric Tons per Capita) | Data," accessed August 18, 2017, https://data.worldbank.org/indicator/.

39. World Bank.

40. Consider the following analogy: Imagine two families land on an island to establish their own respective communities. Both families need to use wood from the forest on the island. Imagining the forest to be practically unlimited, they each just use the wood as they please. The Smiths are industrious and build a sizeable community of log cabins, relying heavily upon the wood in the forest. The Gardeners, devoting more of their time to leisure, have used very little of the forest's wood and are currently all sharing a single cabin, constructed of a small amount of wood from the forest. One day it is discovered that the forest is, contrary to their original estimates, far from unlimited; in fact, it is now nearly depleted. There is a small amount of remaining forest that can be harvested. Given that there are now competing claims to its wood, should the past consumption by the Smiths be completely disregarded, or do we think it a morally relevant factor when deciding upon the fair distribution of the remaining wood?

41. World Bank, "CO_2 Emissions (Metric Tons per Capita) | Data."

42. This is, of course, ignoring the previously mentioned distinction between "subsistence emissions" and "luxury emissions" in Shue, "Subsistence and Luxury Emissions" If we were able to disaggregate subsistence emissions and luxury emissions and only attach moral relevance to luxury emissions, it is doubtful that the per capita emissions of those in the developing world would overtake those of the developed world any time soon.

43. Shue, "Global Environment and International Inequality," 536.

44. Peter Singer, "One Atmosphere," in *Climate Ethics: Essential Readings*, edited by Stephen Gardiner, Simon Caney, Dale Jamieson, and Henry Shue (Oxford: Oxford University Press, 2010), 189.

45. Neumayer, "In Defence of Historical Accountability for Greenhouse Gas Emissions," 189.

46. Neumayer, 189.

47. M. A. Janssen and den Elzen MGJ, "Allocating C02-Emissions by Using Equity Rules and Optimization," *National Institute of Public Health and Environmental Protection – Bilthoven, The Netherlands*, 1992.

48. United Nations Development Program, ed., *Work for Human Development*, Human Development Report 2015 (New York: United Nations Development Programme, 2015), hdr.undp.org/sites/default/files/2015_human_development_report.pdf.

49. This example is taken from Moellendorf, *The Moral Challenge of Dangerous Climate Change*, 169.

50. Edward A. Page, "Give It up for Climate Change: A Defence of the Beneficiary Pays Principle," *International Theory* 4, no. 2 (July 2012): 300–330, https://doi.org/10.1017/S175297191200005X.

51. Page, 316–17.

52. Moellendorf doubts that previous generations' "trespass" on the atmospheric sink amounts to an injustice, even a non-blameworthy one. He notes that whether or not previous generations' use of the atmospheric sink will generate harm for current and future members of the globe depends upon whether or not our current generation decides to take its mitigation obligation seriously. I'm not so sure. By 1990, the atmospheric concentration of carbon dioxide had risen from its 1750 level of 280 ppm to above 350 ppm. It is doubtful that such an increase would not have carried with it any climatic harms whatsoever, even if we had immediately reduced our emissions to zero—a feat that would have been, quite literally, impossible. Furthermore, that increase only measures *atmospheric* carbon dioxide. About 25 percent of carbon dioxide emissions in a given year are taken up by the oceans, thus increasing ocean acidity. This increase in ocean acidity has already resulted in the loss of nearly 30 percent of the Great Barrier Reef, and has undoubtedly generated harms that are already being felt by those who

rely upon the oceans for their livelihood. Given this, it seems plausible to portray previous generations' trespass of the atmospheric sink as in instance of unjust enrichment (since they reaped the benefits and passed burdens along), even if it was a blameless injustice.

53. For another account of how beneficiary responsible may be relevant to geoengineering, see Clare Heyward, "Benefiting from Climate Geoengineering and Corresponding Remedial Duties: The Case of Unforeseeable Harms," *Journal of Applied Philosophy* 31, no. 4 (November 1, 2014): 405–19, https://doi.org/10.1111/japp.12075.

54. International Monetary Fund, "World Economic Outlook Database 2016," accessed August 18, 2017, http://www.imf.org/external/pubs/ft/weo/2016/01/weodata/weorept.aspx.

55. Moellendorf is not advocating the use of the HDI to determine claims to the benefits of burdens of SRM but rather to allocate the costs of climate change mitigation fairly. See Moellendorf, *The Moral Challenge of Dangerous Climate Change*, 176.

56. International Monetary Fund, "World Economic Outlook Database 2016"; United Nations Development Program, *Work for Human Development*.

57. Notre Dame Global Adaptation Index, "Vulnerability Rankings | ND-GAIN Index," accessed August 18, 2017, http://index.gain.org/ranking/vulnerability.

58. Notre Dame Global Adaptation Index.

Chapter Six

Procedural Justice

§1 INTRODUCTION

The previous chapter looked at the normative criterion of substantive justice and determined what would be a fair distribution of the benefits and burdens of stratospheric aerosol injection (SAI). A fair distribution of the benefits and burdens of SAI would be one that heavily favors the least well-off members of our global community. So, in order for us to justifiably grant a geoengineering governance institution the respect it needs to coordinate our action, it must go some way toward satisfying the demands of substantive justice. But the substantive distributive outcome is not all we care about. We also care about how the decisions regarding research, development, and deployment are arrived at. That is, along with substantive justice, a geoengineering governance institution ought to also sufficiently satisfy the demands of procedural justice. The difference between substantive and procedural justice can be understood as the difference between fairness in the result and fairness in the process, respectively.[1] Though, while the two concepts clearly pick out different values, there are few who would consider them completely independent. Most would recognize that any plausible conception of procedural justice must take substantive justice into account, and vice versa. As Rawls remarks, "the justice of a procedure always depends . . . on the justice of its likely outcome, or on substantive justice."[2] So, with the previous chapter focusing on a fair result, the focus of this chapter is on the normative criterion of procedural justice: the fairness of the decision-making process governing SAI.

The main philosophical task of the chapter is to put forward a general principle that can provide some content to the normative criterion of procedural justice used to assess the legitimacy of SAI governance. The kind of general principle we are looking for is one that can provide (a) *fair terms of inclusion*—that is, an answer to the question of *who* ought to be included in the decision-making process—and (b) *fair terms of participation*—that is, an answer to the question of *how* those who ought to be included should be allowed to participate. We need a principle that can answer both the *who* and *how* questions of procedural justice in a way that that all interested parties, if fully informed and motivated to reach agreement, could not reasonably reject. With such a general principle of procedural justice in hand, the second (more applied) task of the chapter is to then identify some real-world implications of relying upon such a principle to design or evaluate an institution charged with overseeing SAI research, development, and deployment.

The chapter is organized into seven sections. The following section will start with an outline of different notions of procedural justice: "imperfect," "perfect," "pure," and "quasi-pure procedural justice." Section 3 then lays out the instrumental and intrinsic value of procedural justice and why it is that we should concern ourselves with fair processes at all. It will be argued that fair processes have both instrumental and intrinsic value. They have instrumental value in that they can secure desirable outcomes, and intrinsic value in that they allow us to both (a) fulfil our natural duties of justice, and (b) relate to one another in a morally valuable way. Section 4 then looks at general principles that can fill out the normative criterion of procedural justice to be used in legitimacy assessments. After surveying the All Affected Principle and the Equal Influence Principle, the section concludes that the Proportionality Principle is the best general principle to fill out the normative criterion of procedural justice. The Proportionality Principle says that decision-making power should be proportional to the claims that individuals have to influence the decision. I argue the Proportionality Principle does the best job of justifiably providing fair terms of inclusion and fair terms of participation for an SAI decision-making process. Section 5 addresses the objection that apportioning decision-making power according to the strength of individual claims is infeasible and thus a poor norm to guide geoengineering governance. I identify two understandings of infeasibility, arguing that the kind of political infeasibility that may aptly apply to the Proportionality Principle is not normatively troubling. The task of section 6 is to work out the implications the Proportionality Principle entails for the real-world decision-making process around SAI. Section 7 concludes the chapter.

§2 PROCEDURAL JUSTICE: PERFECT, IMPERFECT, PURE, AND QUASI-PURE

As was just stated, procedural justice refers to the fairness of a decision-making process. There are three common notions of procedural justice that we would do well to acknowledge: "imperfect procedural justice," "perfect procedural justice," and "pure procedural justice."³ While each of these kinds of procedural justice is appropriate under different circumstances, none of them are a perfect fit for geoengineering governance. Instead, this section concludes that a notion of "quasi-pure procedural justice" should be adopted for SAI decision-making. But before outlining the idea of quasi-pure procedural justice, let's look at three different, yet related, notions—those of imperfect, perfect, and pure procedural justice.

§2.1 Imperfect Procedural Justice

The notion of imperfect procedural justice is marked by situations in which (a) there is an independent criterion for the correct outcome, but (b) no sure-fired procedure to deliver it. The commonly referenced example of imperfect procedural justice is that of a criminal trial. In every criminal trial, there is an independent criterion for the correct outcome: a guilty defendant should be convicted while an innocent defendant should be acquitted. However, while the independent criterion for the correct outcome is known, there is no known procedure that will certainly and unfailingly deliver us to such an outcome. Sometimes, perhaps the vast majority of the time, the trial procedure does, in fact, deliver the desired result—to wit, most of the time innocent defendants are found innocent and guilty defendants are convicted. But it may be impossible to design the legal code and the rules governing criminal proceedings so as to ensure that a guilty defendant is convicted *if and only if* he or she has, in fact, committed the crime of which he or she is accused. This is why we rely upon the trial process to hopefully deliver the correct ruling. Given the imperfect nature of the system, criminal trials are good examples of imperfect procedural justice.

§2.2 Perfect Procedural Justice

Imperfect procedural justice can be contrasted with perfect procedural justice. Perfect procedural justice is marked by situations in which there is not only an independent criterion for a just outcome, but also a sure-fired procedure to deliver this outcome. To illustrate the notion of perfect procedural justice, consider dividing up a chocolate cake. Assume everyone wants as much of the cake as possible, and that no one's claim to the cake is any stronger than anyone else's. Under this scenario, the fair division of the cake would be an equal division. Everyone should receive a slice of cake that is equally the size of any other slice. Thus, the substantively fair division is known ahead of time. Not only is the fair division known ahead of time, but there is also a fool-proof process to deliver such an outcome. As Rawls notes, "the obvious solution is to have one man divide the cake and get the last piece, the others being allowed to pick before him. He will divide the cake equally, since in this way he assures for himself the largest share possible."[4] Cake division is the paradigmatic example of perfect procedural justice since there is both (a) a predetermined outcome to be aimed at and (b) a reliable process to deliver that outcome.

§2.3 Pure Procedural Justice

Pure procedural justice, on the other hand, is the mirror-image or opposite of imperfect procedural justice. Pure procedural justice is marked by situations in which (a) there is *no* independent criterion for determining the just outcome, but (b) there *is* a correct or fair procedure that, when properly followed, makes the outcome similarly correct or fair. Rawls writes,

> This situation is illustrated by gambling. If a number of persons engage in a series of fair bets, the distribution of cash after the last bet is fair, or at least not unfair, whatever this distribution is. I assume here that fair bets are those having a zero expectation of gain, that the bets are made voluntarily, that no one cheats, and so on. The betting procedure is fair and freely entered into under conditions that are fair. Thus the background circumstances define a fair procedure. Now any distribution of cash summing to the initial stock held by all individuals could result from a series of fair bets. In this sense all of these particular distributions are equally fair.[5]

Robert Nozick saw free-market transactions as examples of pure procedural justice. He argued that there is no predetermined or "patterned" state of affairs that can be considered just *ex ante*. Rather, if the right process is followed (namely, a process governed by his principle of justice in transfer), the distribution of resources at the end of the process is just, whatever the distribution happens to be.[6] Some theorists see idealized democratic procedures as examples of pure procedural justice. They note there is no predetermined set of laws that create substantively just outcomes. Citizens thus rely upon the democratic process as a form of pure procedural justice, making the outcome of the process just, whatever the specific laws turn out to be.

But this commitment to pure proceduralism in the democratic process is difficult to justify when majorities exercise their power to bring about substantive outcomes that are unacceptable to minorities, even when such outcomes are brought about in "democratic" ways.[7] This is why many democratic theorists place constraints on the acceptability of outcomes, even when they are the result of a fair democratic process. Thomas Christiano refers to this as "moderate proceduralism," which denotes something along the lines of pure procedural justice with substantive constraints upon the outcome of the procedure—constraints that are generally conceived of as civil rights.[8]

§2.4 Quasi-Pure Procedural Justice

This idea of pure procedural justice with certain substantive constraints that define the range of acceptable procedural outcomes is similar to what Rawls refers to "quasi-pure procedural justice."[9] When substantive justice is ambiguous or admits of a range of options that all could be considered just, a process of quasi-pure procedural justice can be used to generate an acceptable substantive outcome, provided that it falls within the range of what substantive justice mandates. Take Rawls' difference principle, for example. The difference principle requires that the social and economic inequalities of society be arranged so that they are both (a) to the greatest expected benefit of the least well off, and (b) attached to offices and positions open to all under fair equality of opportunity.[10] However, the difference principle clearly does not pick out any particular set of laws as just. Rather, it sets a substantive constraint on the group of laws and other aspects of the basic structure that are to regulate society. There may be various sets of laws that are all compatible with the difference principle, and society uses the legislative process to determine the specific content of those laws, which are considered just as long as they lie within the range of substantive justice identified by the two principles of justice.

Take a non-political example. Imagine a mother telling her three children that they can choose the movie the family will see in the theater this coming Friday. The only constraint being that it must be rated for children of their age. Imagine further that there are two movies currently showing that are rated for children of their age. The children unanimously agree that majority rule will be used to decide which movie the family will see. In this case, neither of the two movies is the one "correct" or "fair" choice. Whatever movie receives two votes is the movie the family ought to end up seeing. The process of majority rule in this example is an instance of quasi-pure procedural justice, since the outcome of the process is considered fair provided that the majority chose one of the two movies that fall within the substantive constraint of being age-appropriate.

Accordingly, it could be said that this chapter relies upon a "quasi-pure" conception of procedural justice as it relates to SAI. We can think of the conclusion from the previous chapter—the idea that the distribution of benefits and burdens associated with SAI should heavily favor the least well-off members of the global community—as a substantive constraint that admits of a range of possible SAI deployment scenarios (including non-deployment). However, this substantive principle does not pick out any particular SAI scenario as just—rather, it identifies an acceptable range of scenarios that all align with the requirements of distributive justice. So, the task of this chapter is to outline a principle of procedural justice that, when applied to the decision-making process around SAI, will generate a just substantive outcome, provided that it falls within the range of outcomes that can be said to heavily favor the least well off. That is, as long as the result of the process is one in which the distribution of benefits and burdens is skewed heavily in favor of the least well-off members of the global community, the outcome can be considered just. But before introducing such a principle and then applying it in practice, we first turn to why it is that we are concerning ourselves with procedural justice at all.

§3 REASONS TO PURSUE PROCEDURAL JUSTICE

One could question why we should even concern ourselves with the idea of a fair process. Especially given the highly technical nature of SAI, why not just appoint a committee of experts to implement the substantive conclusion of the previous chapter? There is controversy about the reasons we have to pursue fair political procedures. Some argue that the reasons we have to pursue fair political procedures are exhausted by the good consequences such procedures tend to produce.[11] For instance, Richard Arneson argues that "democratic procedures, like all procedures, should be evaluated according to the moral value of the outcomes they would be reasonably expected to produce."[12] We can call such a view "pure instrumentalism."[13] Contrary to the pure instrumentalist view, many democratic theorists argue that democracy (being the exemplar of procedural justice) is valuable in and of itself, in addition to the desirable policies it manifests. According to theorists like Thomas Christiano, "the democratic process has an intrinsic fairness." In this section, I argue that we have at least four reasons to concern ourselves with procedural justice when it comes to an institution overseeing climate engineering. Two of these reasons are instrumental reasons, suggesting that an institution guided by norms of procedural justice will deliver certain desirable ends. The other two reasons could be considered intrinsic reasons in that they count in favor of pursuing procedural justice without referencing the good outcomes that fair procedures may engender. Consider first the two instrumental reasons to aim for a fair decision-making process around SAI.

§3.1 Instrumental Reasons to Pursue Procedural Justice

Instrumentally, a fair political procedure can deliver morally desirable outcomes. First, as was mentioned in the previous section, fair procedures can generate a just substantive outcome when that just outcome is not known ahead of time.[14] Our lack of prior knowledge about a just outcome can be due either to (a) the fact that there is no just outcome independent of the process (as in the case of pure procedural justice); (b) the fact that we are simply unable to differentiate between the just outcome and unjust alternatives (as in the case of imperfect procedural justice); or (c) the fact that there is a range of substantively just outcomes that require a selection process to single one out (as in the case of quasi-pure procedural justice). In any of these cases, a fair procedure can have instrumental value insofar as it aids us in establishing, uncovering, or deciding upon a just substantive outcome—or at least an acceptable or agreeable substantive outcome. Along these same lines, fair political procedures often require broad participation, ensuring greater cognitive diversity (which may facilitate better decision making). That is, given that just procedures tend to boast greater inclusivity and can gain input from individuals with a wide range of viewpoints and expertise, this more expansive knowledge base can in turn lead to better substantive outcomes that may have been inaccessible to a more restricted group of decision-makers.

Second, the value of the substantive outcomes decided upon via a just procedure is heavily dependent upon those outcomes being instantiated and adhered to. A wise, knowledgeable, and benevolent dictator could conceivably produce better policy than a large democratic assembly regulated by a consensus decision-making rule. However, if the good policy enacted by the benevolent dictator is rejected by those whose action it is meant to regulate, then the beneficial consequences that such a policy is supposed to carry with it never materialize. There is a significant body of empirical literature suggesting that individuals and states are much more willing to comply with institutional directives when they consider the processes whereby the directives were decided upon to have been guided by norms of procedural justice.[15] That is, when participants bound by a decision-making process recognize the process as just, they are more likely to comply with its outcome even when they are less than fully satisfied with it. This allows a just procedure to secure its substantive outcome more efficiently with less defection.

Both of these considerations are relevant to a decision-making procedure for SAI. First, a geoengineering institution guided by norms of procedural justice can produce good policy. (a) While the previous chapter outlined a fair result of deploying SAI, the fair result is underspecified. An institution with a fair procedure can help us to refine that substantive outcome into something more particular and policy oriented. (b) Given the profound effects and significant uncertainties involved with climate policy and especially with SAI, it will be difficult to craft policy that achieves the agreed upon substantively just outcome. In the face of such uncertainty, the cognitive diversity boasted by an inclusive decision-making body can be an asset. And, along the same lines, a fair process that includes all stakeholders will have greater cognitive diversity about not just scientific uncertainties, but also local particularities, leading to better decisions overall. Secondly, it is no surprise that there are divergent visions of what, exactly, would constitute a substantively just outcome of SAI development and deployment, notwithstanding the conclusion of the previous chapter. And when agreement about that substantively just outcome is reached, there may be various policies that could achieve such an outcome. A fair process provides those whose favored policy did not obtain a reason to accept the outcome and comply with it nonetheless.

§3.2 Intrinsic Reasons to Pursue Procedural Justice

There are clearly instrumental or practical reasons to consider procedural justice as it relates to SAI governance. However, we also have at least two reasons to pursue procedural justice independent of these instrumental considerations. First, we have what Rawls calls a "natural duty of justice" to promote and comply with just institutions. This natural duty is perhaps the most roundly recognized duty among political philosophers. Rawls differentiates obligations, on the one hand, from duties, on the other. Obligations arise from our own voluntary acts and are always owed to definite individuals. For instance, I have an obligation to grade exams. I have this obligation because I voluntarily chose to be an instructor, and the obligation is owed specifically to my students. Unlike obligations, duties "apply to us without regard to our voluntary acts . . . [and] obtain between all as equal moral persons."[16] According to Rawls, our natural duty of justice "requires us to support and to comply with just institutions that exist and apply to us. It also constrains us to further just arrangements not yet established."[17] Thus, we have a natural duty to establish and support just institutions, with the justice of institutions depending both upon substantive and procedural attributes. It is easy to see how this consideration speaks in favor of constraining a geoengineering governance institution with norms of procedural justice. No such institution may yet exist. However, to the extent that we ought to establish such an institution, we have a natural duty to ensure that it complies with the demands of substantive justice and, more pertinent to our present discussion, procedural justice as well.

There is a second intrinsic reason that speaks in favor of pursuing fair political procedures for SAI governance. Fair political procedures—procedures that are governed by terms that all interested parties, if fully informed and motivated to reach agreement, could not reasonably reject—allow us to relate to one another in a morally valuable way. In deciding to regulate our collective decisions on terms that all have reason to accept, we publicly recognize one another as not just equal subjects of the political process, but as equal participants as well.[18] Now, there may be no better paradigm of a global public good than our shared climate system. Unilaterally deciding to manipulate this global public good would exemplify a kind of disrespect for those who would reasonably reject the exclusive terms of such a decision-making process, even if the outcome of the process were to benefit them. Conversely, when making decisions about our shared environment on terms that everyone has reason to accept, we relate to one another as equals who are entitled to not just fair treatment, but fair participation too. This way of relating to one another, as Eric Beerbohm notes, "has immediate and irreducible value."[19]

Thus, we have both instrumental and intrinsic reasons to pursue procedural justice in SAI governance. Instrumentally, just procedures can produce morally desirable outcomes and offer a reasonable justification to—and thus more reliably secure the compliance of—those who would have preferred a different substantive result of the process. In addition to such instrumental considerations, we have a natural duty to promote and comply with institutions governed by norms of procedural justice. And doing so allows us to stand in a relation of mutual accountability to one another. That is, resolving problems of collective decision making through just processes allows us to relate to one another in an intrinsically valuable way, recognizing one another as equal subjects and equal participants of the process. The instrumental value and intrinsic value of just procedures are not mutually exclusive. We can value just procedures in and of themselves, and still be concerned with the results they are likely to produce. And, conversely, we can place significant weight on the outcomes of just procedures while still recognizing that even a full account of the best of outcomes fails to fully capture the entire value of a just procedure.

§4 GENERAL PRINCIPLES FOR PROCEDURAL JUSTICE

Having highlighted the instrumental and intrinsic value of just procedures, in this section I'll assess some general principles that could provide some content to the normative criterion of procedural justice meant to guide an institution overseeing SAI. We need a general principle that provides us with answers regarding fair terms of inclusion and fair terms of participation, with "fair terms of inclusion" referring to *who* ought to be included in the process, and "fair terms of participation" referring to *how* they should be included. We need answers to the *who* and the *how* questions that would be agreeable to all. To be more specific, we need answers to the *who* and *how* questions that all relevant parties, if fully informed and aiming for agreement, could not reasonably reject.

§4.1 The All Affected Principle

The question of who ought to be included in a collective decision is often neglected when theorizing about procedural justice. Generally, theorists think about procedural justice within a pre-given international system, or sovereign state, or polis of some sort. But who should be included in the decision-making process is the first question any conception of procedural justice must answer. Despite it often being neglected, the *who* question has been discussed among a swathe of democratic theorists, referring to it as "the boundary problem" in democratic politics.

Perhaps the reason that the question of fair inclusion has been neglected is that deciding who it is that ought to have a say in any decision is more difficult than it may appear. To use an example from Gustaf Arrhenius, who ought to have a say in the resolution of the conflict in Northern Ireland?[20] Should an answer to the *who* question as it relates to the conflict specify that only the citizens of Northern Ireland ought to have say? Or perhaps residents of the Republic of Ireland ought to be included as well, as they were in the referendums that established the Good Friday Agreement. But why not include the rest of the British citizenry as well? Note that an appeal to democracy does not solve this problem. British and Irish nationalists may be fervent supporters of democracy, yet they will provide different answers to the question of who ought to be involved in determining the proper status of the Northern Irish territory.

There is a well-known standard that has been deemed the best general principle for providing an answer to the boundary problem: namely, the All Affected Principle.[21] Garnering both supporters and critics,[22] the principle expresses the idea that:

> Everyone affected by the outcome of a political decision ought to be included in the decision-making process.[23]

The idea that all those affected by a decision ought to have a say in it has significant intuitive appeal. There are some decisions that appear as though they can be justifiably restricted to a specific group of people, and the All Affected Principle can buttress such restrictions. For example, my home town of Encinitas has an ordinance requiring residential buildings to have a maximum height of two stories (26 feet from the ground to where the roof begins). The residents of Encinitas made this decision to keep a certain aesthetic for the quiet beach community. It seems reasonably justifiable that this kind of decision is left to the residents of the community, perhaps because they are the only ones affected by the outcome of the decision. On the other hand, we don't think the decision of whether or not to allow the dumping of radioactive waste onto the beaches of Encinitas is something that the residents of Encinitas ought to have the exclusive right to decide for themselves. The malign transboundary effects of dumping radioactive waste on Encinitas beaches make the proper scope of inclusion certainly extend beyond merely the residents of the small beach community. And again, the All Affected Principle supports such a conclusion. It would plausibly be an unjustifiable invasion into the autonomy of the people of Encinitas if, say, the U.S. federal government were to intervene and require the city council to allow a residential skyscraper to be built right on the ocean front. And if the people of Encinitas were to claim the exclusive right to decide whether or not dumping radioactive waste on their beaches were to be permitted, others affected by the outcome of the decision could reasonably reject such a claim. The All Affected Principle can provide a normative ground for such conclusions.

Now, while the All Affected Principle may be able to answer the question of who it is that ought to be involved in any political decision—to wit, it can serve as a normative principle underlying fair terms of inclusion—it suffers from a fatal silence on the question of fair terms of participation, or how it is those parties the principle picks out ought to be included in the decision-making process. The principle merely picks out those that ought to be included in the decision-making process, and says absolutely nothing about *how* they ought to be included. Despite its promise regarding fair terms of inclusion, the All Affected Principle cannot serve as a general norm for procedural justice if it cannot fill out the idea of fair terms of participation.

§4.2 Equal Influence in the Decision-Making Process

If we accept that the All Affected Principle accurately picks out the group of those who should be included in the decision-making process, and also accept that there needs to be an answer to the question of how those identified parties ought to be included, perhaps the first answer one would put forward is the idea that everyone should have *equal influence*. Thus, according to what we can call the "Equal Influence Principle":

> Everyone affected by a political decision should have equal influence over the decision-making process.

The Equal Influence Principle is borne out of a conception of political equality that Charles Beitz calls the "Simple View." The Simple View considers the egalitarian commitment to fair terms of participation as an institutional commitment, mandating something like equal decision-making power or equal power to influence the outcome of the process.[24] This is a powerfully intuitive conception of political equality, one that supports our commitment to the idea of "one person, one vote." Now, when legitimate claims to inclusion are equal, the Simple View correctly interprets the egalitarian commitment and requires terms of participation in which equal claims are translated into equal procedural influence. The egalitarian commitment to justify the terms of participation to all parties (conceived of as free and equal autonomous persons) is satisfied. Here, equality at the justificatory level correctly translates into equality at the institutional or procedural level.

Yet, despite its intuitive appeal and success in cases where claims are equal, the Simple View (and thus the Equal Influence Principle) is an inadequately impoverished interpretation of the egalitarian commitment to fair terms of participation. Think again of my home town of Encinitas. My father lives in the city all year round. Consider his claim to participation in the decision-making process surrounding the aesthetic standards for housing compared to someone who vacations in Encinitas for one week each summer. The stakes for my father are significantly higher than they are for the summer vacationer. Providing each with an equal say in the decision-making process would actually be unfair. It would seem to contravene the formal standard of justice that dates back to Aristotle, that of "treat like cases alike," given that the situation of the vacationer and my father are not exactly, or even roughly, alike.[25] Insofar as we are looking for *fair* terms of participation, we need to recognize that fair terms of participation may not necessarily imply *equal* terms of participation. The Simple View confuses fair terms of participation, which obliges us to provide a reasonable justification of the terms of participation to each person with a claim to be included, with institutional equality, which, as mentioned, can be understood as something like equal power to influence the decision. The problem with the Simple View is that it conflates equality at the justificatory level with equality at the institutional or procedural level.[26] When claims to inclusion and participation are unequal, the parity between justificatory and institutional equality breaks down. As Beitz explains, "At the level of institutions, the sovereign regulative ideal is not equality at all but rather fairness."[27]

§4.3 The Proportionality Principle

It seems clear that fair terms of participation cannot translate *unequal* claims into *equal* procedural influence. This would be an affront to the idea of fairness and would be reasonably rejectable from the point of view of those with stronger claims. So, what does fairness imply at the institutional level when claims to participation are unequal? The most plausible interpretation is that fairness requires procedural influence which is *proportional* to claims of participation. If we understand claims of participation as tracking the stakes parties have in the outcome of the decision, we would get something like the Proportionality Principle advanced by Harry Brighouse and Marc Fleurbaey. The Proportionality Principle states:

> Power (or influence) in any decision-making process should be proportional to individual stakes (or claims).[28]

The Proportionality Principle coincides with the Equal Influence Principle in mandating equal power in the decision-making process *if* claims to participation (understood as stakes) are equal. But it departs from the Equal Influence Principle (and the Simple View) once claims to participate diverge, and it requires that procedural influence track the degree to which claims diverge.[29]

When deciding upon the fundamental principles that are to regulate the basic structure of society, it seems reasonable that respect for the equality of persons would translate into procedural equality, since everyone has roughly equal stakes and thus an equal claim to determine such fundamental principles. But not all political decisions are of the same nature as decisions about the fundamental principles that regulate the basic structure of society—think again of my dad and the summer vacationer. Some decisions will have profound impacts upon some, while only mildly affecting others—though, still in a morally relevant way that legitimizes their claim to inclusion.[30] In such instances, we could not expect the person liable to profound impact to accept terms of participation that would apportion them the same influence over the decision as their fellow participant who stands to be only mildly affected. In such a case, apportioning influence in an institutionally equal way actually undermines our egalitarian commitment.

Nearly everyone recognizes that the impacts of climate change will vary significantly. Not only will the impacts of climate change vary significantly, but individuals' ability to respond to these impacts is heavily dependent upon available infrastructure and disposable capital. Thus, in general, a poor agriculturist living in a small island developing state is liable to profound impacts from climate change, whereas an upper-class American living in Colorado is almost certainly less at risk. With these individuals having such radically different stakes in the decisions we have to make about climate change and climate engineering, granting them equal influence over the decision-making process would constitute a kind of procedural unfairness. The Proportionality Principle recognizes the different stakes these individuals have regarding the decisions that need to be made about climate change and climate engineering, and it translates these different stakes into varying degrees of influence over the decision-making process. While still recognizing the ideal of equality at the justificatory level, the Proportionality Principle does a better job of capturing fairness at the institutional level.

§5 FEASIBILITY CONSTRAINTS

The previously outlined principle of procedural justice in which power over political decisions is allotted in proportion to the claims individuals have to that power is certainly not uncontroversial. It nonetheless represents a plausible interpretation of fair terms for a decision-making process, terms that all interested parties, if fully informed and motivated to reach agreement, could accept. To place it within the context of geoengineering governance, the

Proportionality Principle is a plausible interpretation of what the normative criterion of procedural justice demands when it comes to an institution overseeing the intentional manipulation of the planetary environment. But there is a clear problem with the principle. Even if one accepts it as theoretically plausible, it seems regrettably infeasible. It is clear that no such decision-making process in which individuals have power proportional to their claims will govern geoengineering (or any large-scale political decision-making process in the foreseeable future, for that matter). To the extent that we want our conceptions of justice (procedural justice included) to aid us in designing and assessing institutional arrangements in the real world, this may seem like a damning criticism. However, despite the infeasibility of perfectly instantiating such a principle of procedural justice in the decision-making process around geoengineering, there is value in identifying what ideal procedural justice demands.

First, remember that what we are looking for is a general principle that can provide content to the normative criterion of procedural justice to be used in assessing the legitimacy of an institution set up to oversee SAI. And, as was explicated in Chapter 4, legitimacy assessments are not binary. An institution can be more or less legitimate depending upon the degree to which it satisfies the various normative criteria that are appropriate for the kind of institution it is. The institution need not perfectly conform to any of the normative criteria in order to be worthy of the kind of respect we grant it to perform its functions. So, despite it being highly unlikely for a geoengineering governance institution to perfectly instantiate the Proportionality Principle in its decision-making process, the principle nonetheless gives us an ideal to aim at. As Brighouse remarks, "when designing institutions, we choose those that can best realize the principles we've offered, even if they cannot realize them fully."[31]

Second, identifying an ideal principle and the reasons that support it can help us in comparative assessments of institutional arrangements even when such arrangements are not explicitly implementing the ideal principle. Understanding the reasoning that leads to the ideal principle, institutional arrangements can be assessed according to the degree to which they instantiate similar reasoning. For instance, if we recognize that the reasoning leading to the Proportionality Principle is based in the idea of the moral equality of persons, to the extent that certain institutional arrangements fare better or worse in recognizing such equality—even if they are guided by a different principle altogether—we can judge them as being closer or further from what our ideal principle of justice demands.

Third, there are at least two different ways of understanding the claim that a certain principle is infeasible—only one of which is, in fact, a problem for ideal theorizing. To say that a principle is infeasible could, on the one hand, mean that it does not currently enjoy any politically realistic possibility of being implemented. We can call this "political infeasibility."[32] On the other hand, to say a principle is infeasible might mean that it is infeasible even under the most conducive of circumstances. We can call this "theoretical infeasibility." Now, it seems clear that the Proportionality Principle is politically infeasible, but I think the charge of theoretical infeasibility is misplaced. It is not impossible to imagine a scenario in which decision-making power over geoengineering is apportioned according to the stakes that individuals have in the outcome of the decisions. In fact, to cite a common example, the Proportionality Principle seems to be the standard underlying the decision-making process for publicly held companies. Power or influence over the decision-making process is directly proportional to the stakes (shares) one has in the company. This isn't to say that public companies are models of procedural justice, but merely to show that the principle is theoretically feasible. Furthermore, the charge that the Proportionality Principle is politically infeasible is a charge that could be leveled against many principles of justice. For instance, the difference principle is a politically infeasible principle of justice given the current political climate of the United States. But that doesn't mean that the difference principle cannot help guide us in our assessment of the current basic structure of U.S. institutions. Thus, if the Proportionality Principle enjoys *theoretical* feasibility, the charge that it is currently *politically* infeasible is not a damning criticism. Still, we'll want to know what a governance institution guided by the Proportionality Principle would look like.

§6 PROCEDURAL JUSTICE IN PRACTICE: MAKING DECISIONS ABOUT GEOENGINEERING

In this penultimate section I'll explore some possible implications of using the Proportionality Principle to guide the SAI decision-making process. Even if we recognize that the principle is unlikely to be perfectly instantiated in the decision-making process, it can still help us in identifying certain characteristics of an institution overseeing such a process.

§6.1 All Included

When it comes to certain decisions, it will often be difficult to evaluate the claims to inclusion of various parties. To look back at the example of Arrhenius about the territory of Northern Ireland, clearly the Northern Irish are affected in a morally relevant way that legitimizes their claims to inclusion. And the same could be said of the rest of the inhabitants of the island, and perhaps even the rest of the British citizenry as well. But there will be difficult cases along the spectrum of claims that count as clearly legitimate to claims that are clearly illegitimate when it comes to inclusion. Fortunately, evaluating the claims to inclusion of those affected by geoengineering is not quite as difficult.

Stephen Gardiner notes that, by definition, geoengineering will have forceful effects on the climate system. "The climate system," Gardiner writes, "is a basic background condition of human life and social organization on this planet."[33] Affecting that background condition would engender morally relevant effects on the entirety of the world's population both present and future. From the occurrence of floods and droughts, and cyclones and hurricanes, to the price of food and the stability of various governments,[34] altering the climate system will have profound and morally relevant effects on every single person on this planet, whether they are aware of such effects or not. And even if, somehow, someone ends up being unaffected by geoengineering, the fact that they *may* have been affected or that they may have been put at risk entitles them to a say. This gives nearly everyone currently alive and those who will come to live in the near and distant future a legitimate claim to inclusion when it comes to decisions surrounding the controversial technology.

§6.2 Representation

Now, if everyone has a claim to inclusion in the decision-making process around SAI, what does that entail for participation? The fact that everyone has a claim to participate in the decisions pertaining to SAI does not necessarily require the decision-making process to *directly* include everyone. The Proportionality Principle is perfectly compatible with representative forms of decision making. In fact, representatives would be required for those who are unable to effectively represent their own interests, such as children, future generations, and the cognitively disabled.

Incorporating the idea of representation, the principle would imply "all individuals should have their interests effectively represented in proportion to their stakes."[35] The most plausible kind of representation on the international stage comes in the form of state representatives. Ideally, state representatives would deliberate and discuss on behalf of the populations they represent. The well-known problem, of course, is that some state representatives do a decent job of effectively representing the interests of their constituents, while others represent interests that barely correlate with those of the individuals for whom they are supposed to stand. For instance, it could be assumed that the interests of the people of a state like Switzerland are represented decently well during international negotiation, given Switzerland's strong democratic credentials. However, it is doubtful we can say the same for the interests of the people of North Korea. Of course, merely identifying that some states are more effective representatives of their constituents' interests doesn't solve the problem. And here we face a dilemma. On the one hand, including the chosen spokesperson of an unrepresentative state fails to effectively represent the interests of those residing within the state. However, excluding such representatives is no solution either (since then they are not only denied *effective* representation, but *formal* representation as well). Perhaps all we can do here is note that states can be unrepresentative of their population's interests to different degrees. And where we can identify agents who represent the interests of certain state populations to a greater degree than the representative appointed by the state, then such agents could act as fiduciaries for the people in lieu of their state-appointed representative. This is clearly far from an ideal solution. But the issue of poorly represented populations on the international stage is difficult to address, to the say the least.

§6.3 Populations

Another foreseeable implication of the Proportionality Principle is that it would require a stark break with the current international decision-making process of "one country, one vote." The idea of one country one vote places strong weight on the idea of sovereignty. In contrast, the Proportionality Principle may imply that state representatives have decision-making power that is proportional to the size of the population they represent, and thus it may go much further in recognizing the moral equality of persons. Even if we were to assume that the stakes for each individual were the same, how could we expect individuals residing in China (the world's most populous nation) to accept terms of participation in which their representative had the exact same decision-making power as the representative of Tokelau (the world's least populous nation)?[36] In guiding the decision-making process around SAI by the Proportionality Principle, population size would have to be considered.

§6.4 Strength of Claims

The Proportionality Principle implies not just that representatives wield decision-making power in proportion to the population they represent, but also according to the strength of the claims of those populations. Nothing yet has been said about how we should consider the strength of claims to participation when it comes to decisions about SAI. Remember that the Proportionality Principle says that power (or influence) in any decision-making process should be proportional to individual claims. Claims to participation were equated with the legitimate stakes one has in the outcome of the decision. But this immediately invites the question of how to measure the legitimate stakes people have in the decisions surrounding SAI.[37] There will certainly be disagreement in the real world about how to determine what is legitimately at stake for each participant. Different metrics for legitimate stakes could be put forward.[38] But as a rough approximation, perhaps the most plausible way of understanding the stakes that individuals have in the decisions surrounding SAI is to think of their climate vulnerability. Insofar as someone is more vulnerable to climate change, their stakes in decisions about SAI will be greater. We could again rely upon the Notre Dame Global Adaptation Index to determine the vulnerabilities of individuals within countries. Of course, the vulnerability of each individual will differ, but multiplying the country's average vulnerability score by their population will achieve the same result as adding up all individual vulnerabilities (assuming that the index is an accurate representation of the country average). Understanding claims in this way, what we would come out with is a procedure whereby each state representative's decision-making power would be weighted according to the population they represent multiplied by the average vulnerability of that population.[39]

§6.5 Voting

Finally, the Proportionality Principle implies that decisions will most likely have to be made through voting. And recognizing that influence in the decision-making process should track the strength of the claims of the parties involved sheds light on the appropriate *choice rule*. Choice rules determine the formal process used to reach a particular substantive outcome when agreement about that outcome is lacking. At one end of the spectrum lies unanimity, at the other lies decision by a simple majority. Unanimity is an unacceptable choice rule for at least the following three reasons. First, unanimity fails to recognize the genuine disagreement about the best substantive outcome of a decision-making process. We are interested in procedural justice precisely because unanimous agreement about the substantive end is lacking. Thus, in the face of genuine disagreement about substantive outcomes, unanimity is a non-starter. Second, given that every representative essentially has a veto power over the issue being voted upon, unanimity has a built-in status quo bias, allowing those who prefer the status quo to withhold their assent and get their way. This gives them unfair influence over the process. Third, unanimity accords the same influence to all parties involved, regardless of their legitimate claims to participation. There is surely no reason to allocate influence in the decision-making process according to the claims of participants if all decisions must be made unanimously. For at least these three reasons, unanimity is a reasonably rejectable choice rule for collective decision making guided by the Proportionality Principle.

Majority rule, on the other hand, avoids all three of the previously mentioned objections. Now, majority rule has historically been susceptible to the pernicious problem of the "tyranny of the majority,"[40] or the potential for the majority to exercise their power in bringing about substantive outcomes that are unacceptable to minorities. However, while the tyranny of the majority is a real concern for traditional majority rule when accompanied by equal power in decision-making, majority-rule decisions guided by the Proportionality Principle are actually able to prevent majorities from unjustifiably imposing their will on minorities. The fact that simple majority rule will sometimes lead to the violation of basic rights of the minority is what leads to the moderate proceduralism advocated by Christiano that was mentioned in section 2 of this chapter. But the recognition of basic rights and liberties of a minority can be accounted for by the Proportionality Principle, since the greater stakes of those who stand to have their basic rights violated will translate into greater influence in the decision-making process, which will curtail the "majority's" ability to deprive the "minority" of their vital interests.[41]

There may indeed be other prominent characteristics of a geoengineering governance institution guided by the Proportionality Principle. But these five suggestions provide a good starting point. The Proportionality Principle would imply that everyone should have some say over the decisions we face with respect to SAI. The inclusion that all are entitled to could, for practical purposes, translate into everyone being effectively represented, either by a state actor or an agent better positioned to do so. The power each representative would wield would be proportional to (a) the number of people she is representing, and (b) the strength of the claims of those individuals. Finally, the representatives would gather and deliberate about the decisions we face with respect to SAI. After deliberation and discussion, a vote would be held, with each representative's vote being weighted according to the power they ought to wield.

These are all just suggestions of how the Proportionality Principle could be instantiated in a governance institution tasked with overseeing geoengineering. I am under no illusion that the Proportionality Principle would be adopted or perfectly manifested by a geoengineering governance institution. And I don't mean to argue that the above-mentioned considerations are perfectly ideal for geoengineering governance. Still, by recognizing the importance of the Proportionality Principle, we will be able to judge certain decision-making procedures as closer or further from what procedural justice demands. If we are convinced by the idea that the world's most vulnerable populations should have a significant say over if, how, and when geoengineering is to be pursued, then perhaps we could grant veto power to particular representatives. One thing is certain: however decisions are made about the intentional manipulation of the planetary environment, the world's most vulnerable populations should have a central voice.

§7 CONCLUSION

This chapter started off by noting that, along with substantive justice, we also want geoengineering governance to sufficiently satisfy the normative criterion of procedural justice. But, to requote Rawls, it was acknowledged that "the justice of a procedure always depends . . . on the justice of its likely outcome, or on substantive justice."[42] Remember that the kind of procedural justice we pursued in this chapter was what Rawls calls "quasi-pure procedural justice," implying that the outcome of the procedure would be just, provided that it falls within the range of just substantive outcomes. The thought is that decisions about SAI can be considered just if they are the product of a procedure guided by the Proportionality Principle. However, that outcome would have to fall within the substantive constraint identified in the previous chapter—the benefits and burdens of any decision would have to heavily favor the least well-off members of the global community. So, is an SAI decision-procedure guided by the Proportionality Principle likely to meet the demands of substantive justice? If we assume that the representatives of the least well-off members of the global community are effective at actually representing the interests of their constituents, the fact that these representatives would have the greatest influence over the decision-making process should tend to make the outcome of the process coincide with the demands of substantive justice. Of course, this is not guaranteed. But such a process seems likely—and perhaps more likely than available alternative processes—to adhere to what substantive distributive justice requires.

NOTES

1. This is how the difference is described by John Rawls, *Political Liberalism* (New York: Columbia University Press, 2005), 421, and Stuart Hampshire, "Liberalism: The New Twist," *New York Review of Books*, August 12, 1993.
2. Rawls, *Political Liberalism*, 421.
3. John Rawls, *A Theory of Justice* (Cambridge, MA: Belknap Press of Harvard University Press, 1999), 74–75. The following examples are taken from Rawls as well.
4. Ibid., 74.
5. Ibid., 75.
6. Robert Nozick, *Anarchy, State, and Utopia* (Malden, MA: Blackwell, 2012), Chapter 7. This is, of course, assuming that the distribution prior to the transfers was also just, which is covered by Nozick's principle of "justice in acquisition."
7. I put "democratic" in scare quote because many proceduralists will maintain that truly fair and democratic procedures could not lead to outcomes that jeopardize minority rights. See Rawls, *Political Liberalism*, 423.
8. Thomas Christiano, *The Constitution of Equality: Democratic Authority and Its Limits* (Oxford: Oxford University Press, 2008), 296.
9. Rawls, *A Theory of Justice*, 318.
10. Ibid., 266.

11. Richard Arneson argues along these lines in Richard Arneson, "Democracy Is Not Intrinsically Just," in *Philosophy, Politics and Society*, ed. Peter Laslett and James Fishkin, 5 (New Haven, CT: Yale University Press, 1979).

12. Richard Arneson, "Democratic Rights at the National and Workplace Levels," in *The Idea of Democracy*, ed. David Copp, Jean Hampton, and John E. Roemer (New York: Cambridge University Press, 1995), 125, as cited in Christopher G. Griffin, "Democracy as a Non–instrumentally Just Procedure," *Journal of Political Philosophy* 11, no. 1 (2003): 111–121.

13. Griffin, "Democracy as a Non–instrumentally Just Procedure."

14. Rawls, *A Theory of Justice*, 75.

15. Tom R. Tyler, *Why People Obey the Law* (New Haven, CT: Yale University Press, 1992), cited in Kristina Murphy and Tom Tyler, "Procedural Justice and Compliance Behaviour: The Mediating Role of Emotions," *European Journal of Social Psychology* 38, no. 4 (June 2008): 652–68.

16. Rawls, *A Theory of Justice*, 98–99.

17. Ibid.

18. The idea of relating to one another as both participants and subjects—or as co-authors and co-subjects—is similar to what Rainer Forst calls "political" and "legal autonomy," respectively. See Rainer Forst, *The Right to Justification: Elements of a Constructivist Theory of Justice*, trans. Jeffrey Flynn (New York: Columbia University Press, 2012), 128–37.

19. Eric Beerbohm, *In Our Name: The Ethics of Democracy* (Princeton, NJ: Princeton University Press, 2015), 37.

20. Gustaf Arrhenius, "Democracy for the 21th Century," *Sociology Looks at the Twenty-First Century*, 2013, 2–3.

21. Robert Dahl refers to the All Affected Principle as the best standard for determining the demos. See Robert A. Dahl, *After the Revolution; Authority in a Good Society* (New Haven, CT: Yale University Press, 1970), 64.

22. Also labeled "the problem of defining the demos," the boundary problem in democratic politics has been taken up by many. See, for instance, Gustaf Arrhenius, "The Boundary Problem in Democratic Theory," *Democracy Unbound: Basic Explorations I*, 2005, 14–29; Robert E. Goodin, "Enfranchising All Affected Interests, and Its Alternatives," *Philosophy & Public Affairs* 35, no. 1 (2007): 40–68; Robert E. Goodin, "Enfranchising the Earth, and Its Alternatives," *Political Studies* 44, no. 5 (1996): 835–849; Dahl, *After the Revolution; Authority in a Good Society*; David Owen, "Constituting the Polity, Constituting the Demos: On the Place of the All Affected Interests Principle in Democratic Theory and in Resolving the Democratic Boundary Problem," *Ethics & Global Politics* 5, no. 3 (January 2012): 129–52; Sofia Näsström, "The Challenge of the All-Affected Principle," *Political Studies* 59, no. 1 (March 2011): 116–34; Clare Heyward, "Can the All-Affected Principle Include Future Persons? Green Deliberative Democracy and the Non-Identity Problem," *Environmental Politics* 17, no. 4 (August 2008): 625–43.

23. For a similar formulation of the principle, see Robert A. Dahl, "Procedural Democracy," in *Philosophy, Politics and Society*, ed. Peter Laslett and James Fishkin, 5 (New Haven, CT: Yale University Press, 1979).

24. Charles R. Beitz, *Political Equality: An Essay in Democratic Theory* (Princeton, NJ: Princeton University Press, 1989), 4.

25. This idea of justice requiring that we treat like cases alike is often attributed to Aristotle. See Aristotle, *Nicomachean Ethics*, Book V, Chapter 3 1131a10-b15.

26. Beitz, *Political Equality*, 16.

27. Ibid., 18.

28. Harry Brighouse and Marc Fleurbaey, "Democracy and Proportionality," *Journal of Political Philosophy* 18, no. 2 (June 2010): 137–55. Brighouse and Fleurbaey flesh out the idea of stakes in terms of interests, "where interests are conceived broadly and not exclusively in terms of subjective preferences or states" (p. 138).

29. This view is compatible with the conception of political equality advanced in Ronald Dworkin, *Justice for Hedgehogs* (Cambridge, MA: Belknap Press of Harvard University Press, 2011), 388–92. Dworkin writes, "If the legitimacy of a political arrangement can be improved

by constitutional arrangements that create some inequality of impact but carry no taint or danger of indignity, then it would be perverse to rule these measures out. That is the fatal weakness of the majoritarian conception. It rightly emphasizes the value of equal impact, but it misunderstands the nature and hence the limits of that value; it compromises the true value at stake, which is positive liberty, by turning equality of impact into a dangerous fetish."

30. Nozick has a wonderful example of showing what kind of claims are morally irrelevant and thus illegitimate. He asks us to imagine a woman with four suitors. The woman's decision about whom to marry—or, indeed, whether to marry or not—will potentially have significant effects on all five individuals involved, not to mention others who will also be impacted by which of the suitors are still available after her decision. But no one would maintain that all five individuals—and certainly not the more inclusive group of all those potentially affected—ought to have a say in whom, if anyone, the woman shall marry. This shows that not all claims to inclusion are legitimate. See Nozick, *Anarchy, State, and Utopia*, 269.

31. Harry Brighouse, *Justice*, Key Concepts (Cambridge: Polity, 2004), 19.

32. Ibid., 27.

33. Stephen Gardiner, "Is Arming the Future with Geoengineering Really the Lesser Evil?," in *Climate Ethics*, ed. Stephen Gardiner et al. (Oxford: Oxford University Press, 2010), 294.

34. Nick Obradovich, "Climate Change May Speed Democratic Turnover," *Climatic Change* 140, no. 2 (January 2017): 135–47.

35. Brighouse and Fleurbaey, "Democracy and Proportionality," 150.

36. As of 2017, China had a population of roughly 1.4 billion, whereas Tokelau had a population of around 1,300.

37. In an idealized scenario, people would deliberate and reach agreement about the content underlying each others' claims and thus determine the strength of claims as well.

38. For instance, we could understand stakes in economic terms by looking at the expected financial loss individuals can expect from climate change. Or we could understand stakes in terms of rights violations.

39. For a similar approach to modifying the UNFCCC process (though, also including concerns of mitigation), see Luke Kemp, "Framework for the Future? Exploring the Possibility of Majority Voting in the Climate Negotiations," *International Environmental Agreements: Politics, Law and Economics* 16, no. 5 (October 2016): 757–79.

40. John Stuart Mill, *On Liberty and Other Essays*, Oxford World's Classics (Oxford: Oxford University Press, 2008), 8.

41. Brighouse and Fleurbaey, "Democracy and Proportionality," 140.

42. Rawls, *Political Liberalism*, 421.

Chapter Seven

Conclusion

§1 INTRODUCTION

In the previous six chapters, I've attempted to make some headway toward addressing normative issues raised by the prospect of engineering the climate. I argued that research should go forward, that deployment may not be unjustifiable, and that we should regulate geoengineering with a legitimate governance institution guided by, among other things, norms of substantive and procedural justice. But even if I have made some headway addressing these normative issues, I have also most certainly fallen far short of addressing every concern about climate engineering, let alone addressing every concern adequately. Before concluding this book, I'd like to briefly outline some of the remaining complexities a technology like stratospheric aerosol injection (SAI) raises. I want to briefly say something about (1) our non-ideal situation, (2) intergenerational justice, and (3) anthropocentric ethics.

§2 OUR NON-IDEAL SITUATION

Take a second to return to 1988. James Hansen has just testified in front of the U.S. Congress, proclaiming: "Global warming has begun." We know now that Hansen's projections of what was to come were astoundingly accurate.[1] If we had known then what we know now, what would have been the ideal response?

In 1988, atmospheric concentrations of greenhouse gases were at 410 ppm, and yearly global greenhouse gas emissions were roughly 35 Gt of CO_2eq.[2] If we had started mitigating global emissions immediately, the curve toward net zero emissions by 2050 would have been rather gradual. We could have started with five years or so of simply stagnant global emissions.

We then could have transitioned to decades of slowly declining emissions at, say, 2–3 percent decline per year. Eventually, after renewables had actually become more cost-efficient, we could have arrived at net zero emissions. And this all could have been done while making sufficient investments in adaptation projects to address any of the climate change that had been locked in by historical emissions. This would have been a kind of ideal scenario for addressing climate change. I certainly don't want to imply that this would have been *the* ideal scenario, and that anything that deviated from this would have been suboptimal. There are a number of different scenarios that could have achieved similar results. Still, this would have been a great response to the problem at hand.

Unfortunately, the path we've actually followed has deviated drastically from the one just outlined.[3] Remember that global GHG emissions in 1988 were at roughly 35 Gt CO_2eq. Thirty years later in 2018, rather than having decreased, global GHG emissions have risen to over 50 Gt CO_2eq, an increase of over 40 percent. As of 2017, the National Oceanic and Atmospheric Administration measured atmospheric greenhouse gas concentrations to be at 493 ppm.[4] According to the IPCC, atmospheric concentrations of greenhouse gases need to stabilize at roughly 450 ppm if we are to likely limit warming to 2°C.[5] Thus, if we need atmospheric GHG concentrations to stabilize at 450 ppm and last year we were at 493 ppm, we may already be committed to relying heavily upon carbon dioxide removal (CDR) technologies.

At the request of the parties to the UNFCCC, the IPCC just released its special report on limiting climate change to 1.5°C.[6] Every scenario they present that is consistent with limiting warming to 1.5°C relies upon CDR technology. The problem with relying upon CDR technology is that its scalability is uncertain; no one knows whether the technology can be scaled up to the magnitude envisioned, especially in the short time frame available. Take bioenergy with carbon capture and storage (BECCS), for example. The bonus of BECCS facilities is that they create energy while also capturing carbon emissions. To put things into perspective, there are currently three BECCS power plants in use. According to research by Lenzi et al., these three BECCS facilities would have to grow to 160,000 by 2050 in order to capture the amount of CO_2 required by some of the IPCC scenarios.[7] And this investment in research and development of CDR technologies would have to happen alongside emissions decreasing at an unprecedented rate. Carbon Brief has put forward a scenario in which we likely limit warming to 2°C, but such a scenario requires global emissions to drop by 5 percent each year until 2023 and then drop by 9 percent every year for the next decade.[8] These are emissions reductions that some contend are incompatible with a growing economy.[9]

All of this is merely to say that the scenario in which we currently find ourselves is far from ideal. It is this far-from-ideal scenario in which recommendations about SAI are being made. I doubt many would have advocated for research into SAI in 1988 (myself included). The window to address climate change through conventional measures was still wide open. But, in 2018, there was significant uncertainty as to whether avoiding catastrophic climate change was still attainable through mitigation and adaptation alone. The window is certainly being closed, if it is still open at all.

It may be the case that CDR and SRM technologies will not work or that they won't be feasible on the scale needed. This has led scholars like Henry Shue to consider CDR technology to be an unjustifiable gamble—unjustifiable since the gamble is one in which current generations are gambling on behalf of future generations, and future generations are the only ones that can feel the loss.[10] There are significant intergenerational justice concerns at stake. But right now, given the dim prospects of sticking with mitigation and adaptation only, continued research and development of these technologies appears to be the safer bet if one is thinking about future generations. It seems better to leave more options on the table than fewer, especially given the significant uncertainties involved. Indeed, this is the thought behind Rawls' "principle of postponement," that says, roughly: if in the future we may want to do various things—that is, mitigate, adapt, remove carbon from the atmosphere, reflect incoming sunlight—then, other things equal, the action we take now should leave these different options on the table.[11] The reasoning behind the principle of postponement seems right. We don't know whether we will want to use SRM. But it is better to continue research and have the option on the table should it prove useful in the future. The situation would be different if it were still 1988.

§3 INTERGENERATIONAL JUSTICE

The topic of the previous paragraph—intergenerational justice—is something I haven't addressed in any detail throughout the book. This is partly because I agree with Stephen Gardiner that our current political and ethical theories run into great trouble when it comes to thinking about justice at the intergenerational level.[12] This makes arriving at sound normative judgments about integrational issues particularly difficult. Entire books have been written on the topic, and so all I hope to do here is outline some of the issues and highlight their relevance to climate engineering. Hopefully, in the near future, there will be more thought put into the ramifications that climate engineering technologies pose for the project of intergenerational justice.

First, I want to make clear what we are referring to when we talk about intergenerational justice. Traditionally, discussions of justice have been confined to discussions about the competing claims between contemporaries. More specifically, prior to Rawls' discussion of the topic in 1971, justice had often been narrowly conceived of as a proper balance between the competing claims of individuals and groups within overlapping generations. But since the publication of *A Theory of Justice*, there has been a rich body of literature devoted to theorizing about justice between noncontemporaries; individuals and groups within non-overlapping generations. But what is it that we owe future generations? What does justice demand we do for those who will be born in the distant future?

To be sure, there are some who maintain that we have no justice-based obligations to future generations.[13] But to most people, this seems implausible. Granting that we do, in fact, have justice-based obligations to future generations, the question is: What are those obligations? What does justice demand in an intergenerational setting? There are myriad ways one can answer that question, but consider first a utilitarian answer.

§3.1 Utilitarian Intergenerational Justice

Utilitarianism is perhaps one of the strongest moral theories developed to date. While some associate utilitarianism only with moral theory, even one of the founding fathers of the theory—John Stuart Mill—recognized the importance of justice.[14] Utilitarianism, in the abstract, says that actions are right in so far as they produce the greatest amount of net utility (with utility being happiness, pleasure, preference satisfaction, or perhaps rational preference satisfaction). In my mind, the most pressing criticism one can make of a utilitarian theory of justice relates to its disregard for the distribution of utility.[15] When it comes to justice between contemporaries, utilitarianism has been criticized for allowing utility losses for some to be offset by a sufficient quantity or quality of utility gains by others. For instance, if Policy A were to decrease the welfare of 100 people by 10 units each, this may not condemn the policy on utilitarian grounds. If Policy A were to also increase the welfare of 100 other people by 10.1 units each, then the policy would be justified. And, of course, there is a whole spectrum of other scenarios that would also justify Policy A. For instance, if it were to increase the welfare of 10 other people by 100.1 units, this would also justify the decrease in welfare for our original 100 people. And if the policy were to increase the welfare of one single individual by 1,000.1 units, this, too, would justify imposing the loss of 10 units of welfare on the original 100 people. Worse yet, even policies that require huge losses for some group can be justified as long as there is a large enough group that gains something from such a policy. For example, if Policy B were to provide 1,000 people with an increase of only 1 unit of welfare, such an increase could be said to justify imposing a welfare loss of 999 units on a single individual. As long as the policy produces a net increase in utility, utilitarianism supports it. It is exactly these kinds of sacrificial scenarios that led Rawls to say: "Each person possesses an inviolability founded on justice that even the welfare of society as a whole cannot override. For this reason justice denies that the loss of freedom for some is made right by a greater good shared by others."[16]

And the problem with disregarding the distribution of utility when it comes to *inter*generational justice is parallel to the problem of disregarding distribution with respect to *intra*generational justice. Insofar as utilitarians are committed to a *long-termist* view on the consequences of actions, the welfare of one generation could be sacrificed even if it produces only a miniscule increase in welfare across a number of future generations. Imagine that we are considering deploying SAI this year. And imagine that such a deployment would carry with it catastrophic consequences for the next thirty or forty years while we worked out all the possible kinks. Imagine further that SAI would eventually stabilize the climate for a much larger number of generations (with presumably even larger populations) stretching into the distant future. Insofar as we are committed to assessing policies based upon their consequences stretched across a long time horizon, such a deployment of SAI could be justified. But sacrificing one generation for the benefit of others seems as repugnant as sacrificing one individual for the sake of the whole. In fact, it seems even more repugnant if we are to (reasonably) assume that future generations will be better off than previous generations.

§3.2 Sufficientarian Intergenerational Justice

The problems with utilitarianism have led many to look elsewhere for a conception of intergenerational justice with more palatable implications. One of the most plausible conceptions of intergenerational justice is a *sufficientarian* conception. A sufficientarian conception of justice requires that we distribute resources so that everyone meets some minimum threshold of sufficiency, and that inequalities above that threshold can be tolerated. Thus, with respect to intergenerational justice, a sufficientarian conception would require that we leave or pass along enough resources to future generations so that they, too, are able to enjoy this minimum level of sufficiency. The first systematic expression of a sufficientarian conception of intergenerational justice comes again from Rawls.

Rawls split his theory of intergenerational justice into two stages: an *accumulation stage* and a *steady-state stage*. According to Rawls, the minimum threshold that should be considered sufficient is whatever it takes to establish and maintain a just basic structure of society. At any time prior to which society has established a fully just basic structure, this is considered the accumulation stage. Once society reaches a point at which the basic structure is just, we are then in the steady-state stage.

When in the steady-state stage, all that intergenerational justice requires, according to Rawls, is that current generations pass along enough resources for future generations to maintain the just basic structure. But in the accumulation stage prior to the establishment of fully just institutions, a different principle for intergenerational justice applies: the *just savings principle*. The just savings principle requires current generations to set aside resources for future generations at a particular rate. The particular rate at which generations have to save, as with just about everything in Rawls' theory, is determined by Rawls' contractualist method of justification: the original position.

The original position is a hypothetical situation in which parties come together to agree upon principles of justice. Of course, with full knowledge of their specific characteristics, parties are likely to advance principles that would favor people with their particular characteristics. Thus, Rawls has the parties reason from behind the *veil of ignorance*. The veil of ignorance hides from the parties contingencies such as their place in society, their class, their natural abilities such as intelligence and strength, their conception of the good, and even their particular psychological dispositions such as aversion to risk and tendency toward optimism or pessimism. And Rawls adds one's generation to this list, and then asks parties to agree upon a rate of savings "subject to the further condition that they must want all *previous* generations to have followed it."[17]

So, according to Rawls' sufficientarian conception of intergenerational justice, we are not required to make inordinate sacrifices in order to produce miniscule benefits for future generations. Rather, we are required to set aside a certain percentage of resources so that future generations can continue to strive for a just basic structure. While this seems clear enough, what isn't clear is what the just savings principle implies for climate engineering. Perhaps one implication is that we, the current generation, could be asked to make some sacrifices in order to help future generations secure just institutions. For example, perhaps we are required to invest in research so that future generations can help reduce some of the injustice associated with anthropogenic climate change. Or, on the other hand, perhaps we are required to forgo engineering the climate and incur some burden so as to help future generations make strides toward more just institutions. It really isn't clear what intergenerational justice demands with respect to climate engineering, and the previous two vague suggestions are clearly not arguments. What is clear is that more research is needed to determine what kind of constraints and permissions are in play with respect to intergenerational justice when it comes to climate engineering.

§4 ANTHROPOCENTRIC ETHICS

The final topic I want to highlight in the concluding chapter is *anthropocentric ethics*. Within the environmental ethics literature, significant ink has been spilled in attempting to correctly identify the proper scope of moral standing. Nearly all of the analysis in the preceding chapters has focused on anthropocentric issues—issues with humans as the kind of things that ultimately are of moral concern. But, as was briefly alluded to in Chapter 3 during the discussion on respect for nature, one need not give homo sapiens an exclusive seat at the moral table.

Peter Singer has a useful metaphor that he calls the expanding circle. Morality, according to Singer, arose from an evolutionary disposition to protect one's family and close-knit community.[18] Over time, the circle that started with us looking out for ourselves and our friends and family expanded to include others in our tribe, our city, our nation, and ultimately to all of humanity. But many environmental ethicists argue that the circle should continue to expand—that it should expand so as to include the interests of: (a) all sentient creatures; (b) all life; or even (c) all ecosystems.[19]

§4.1 The Expanding Circle

Singer is perhaps the most prominent supporter of drawing the lines of the moral community so as to include all sentient life rather than just humans. Noting that previous liberation movements—the abolitionist movement, the women's liberation movement, the movement for LGBTQ rights, etc.—had expanded the circle of who was due equal consideration, Singer advocates that we keep going. He writes, "I am urging that we extend to other species the basic principle of equality that most of us recognize should be extended to all members of our own species."[20] And Singer's position is not a new one. The founder of the utilitarian tradition, Jeremy Bentham, is famous for saying that, when it comes to the moral considerability of animals, "The question is not, Can they reason?, nor Can they talk? but, Can they suffer?"[21] It is exactly this focus on the ability to experience pain and pleasure that leads Singer to say that sentience is the only rational boundary at which to draw the line of moral considerability.[22] Given climate engineering's profound potential to affect nonhuman animals, those who focus on animal ethics should be concerned.

Singer's proposal to expand our circle of moral considerability has had a huge impact throughout the world. But there are some who maintain that the circle needs to be expanded even beyond the call put forth by Singer. It isn't just sentient creatures that need to be taken into account, it's *all life* that deserves moral consideration. Recall Paul Taylor's biocentrism that was outlined in Chapter 3. He argues that there is inherent value in teleological systems, and that all forms of life—humans, nonhuman animals, plants, etc.—are teleological systems. To reiterate, Taylor's position entails that we have obligations not just *with regard to* nonhuman life, but *to* nonhuman life. For example, if we wantonly destroy a shrub in the middle of Colorado, this act is wrong not because it is something we couldn't justify to our fellow humans.[23] Such an act is wrong because we have failed to fulfill an obligation to respect the shrub itself.

The proposal to respect all life, sentient or not, is rather radical. But others argue that the boundary of moral concern be stretched yet further. Writing in the remote wilderness of Wisconsin, Aldo Leopold argues that humans needed to consider the effects their actions would have not just on animals, nor merely on different plant species, but on the land in its entirety. Developing what he would call *the land ethic*, Leopold argues, "a thing is right when it tends to preserve the integrity, stability and beauty of the biotic community. It is wrong when it tends otherwise."[24] And Christopher Stone pushes the land ethic even further. "I am quite seriously proposing," Stone writes, "that we give legal rights to forests, oceans, rivers, and other so-called 'natural objects' in the environment—indeed, to the natural environment as a whole."[25]

§4.2 A Broader Anthropocentrism

In previous chapters, we saw that SAI has the potential to assuage some of the risks associated with anthropogenic climate change. For instance, SAI could lessen the increase in average global surface temperature, thus providing some breathing room for humans, nonhuman animals, and plants and ecosystems that are acutely sensitive to heat increase. If we were following Leopold's rule—that something is right insofar as it preserves the integrity of the biotic community—such a use of SAI may fall into the category of morally permissible (or even obligatory) actions, given our current situation. But, remember the second part of Leopold's rule: "It is wrong when it tends otherwise."

We know that along with a potential to bring about great good, SAI also has the potential to exacerbate some of the ills associated with anthropogenic climate change. For example, we know that climate change threatens the biodiversity of both animal and plant species. But a SAI program gone wrong could prove even worse. Biologist Christopher Trisos and colleagues have modelled the potential impact SAI could have on biodiversity if it were implemented and then abruptly halted. Relying upon integrated assessment models that incorporate both temperature and precipitation change and the ability of various species to migrate away from temperature and precipitation anomalies, Trisos et al. conclude that "rapid geoengineering termination would significantly increase the threats to biodiversity from climate change."[26]

It should come as no surprise that a sudden and rapid termination of a large-scale geoengineering program would engender substantial harms for humans, nonhuman animals, and entire ecosystems alike.[27] But if geoengineering were to be either bad or good for all communities that deserve moral consideration, then we could hope for what Bryan Norton has termed "the convergence hypothesis."[28] The convergence hypothesis states that, despite having radically different rationales, anthropocentrists, animal ethicists, biocentrists, and ecocentrists might actually converge on similar environmental policy recommendations. But Svoboda has cast doubt upon the convergence hypothesis when it comes to implementing geoengineering.[29] He has shown that there are plausible scenarios in which geoengineering could be used to assuage the effects of climate change for one community—the human community—while doing nothing for or even exacerbating the effects of climate change for other communities. It is interesting to think about what we should do when geoengineering would preserve the integrity, stability, and beauty of one community, but undermine others. How do we go about making decisions when SAI may preserve the integrity of, say, human communities, yet seriously harm oceans, mountains, and the plants and animals that occupy them?

It can be difficult to reach sound normative judgments when dealing with situations in which the interests of multiple communities conflict. Building upon the ecocentric land ethic of Leopold, J. Baird Callicott argues that what we need are hierarchically ordered principles. Leopold's principle, the thought that something is right if it preserves the integrity of the biotic community, is the first principle. But when we have obligations to preserve some parts of the biotic community that clash with obligations to preserve other parts, Callicott offers two second-order principles to deliver a conclusion. "The first second-order principle (SOP-1) is that obligations generated by membership in more venerable and intimate communities take precedence over those generated in more recently emerged and impersonal communities."[30] Thus, with respect to climate engineering, our obligations to our fellow humans would take precedence over our obligations to nonhuman animals, plants, and other parts of the earth's ecosystems. But, "the second second-order principle (SOP-2) is that stronger interests . . . generate duties that take precedence over duties generated by weaker interests."[31] Thus, if the interests of nonhuman animals and plant species that were at stake with respect to climate engineering were to be significantly greater than the interests of our fellow humans, we ought to side with the stronger interests rather than the closer community.

Callicott's ethic of tiered principles is certainly an improvement on Leopold's beautiful, yet vague, prose. But it is doubtful that even Callicott's formulation of the land ethic will be able to deliver a justifiable conclusion in every situation we could face with respect to climate engineering. The idea of "more venerable communities" seems clear enough. But how do we go about totaling up the myriad "interests" of the various communities at stake in SOP-2? Is there really a particular amount of harm that could be done to a grassland in Patagonia that would trump the lives of 10, or 1,000, or 100,000 human individuals? It is clear that we need to broaden our ethical thinking beyond a narrow anthropocentrism. But how much broader it needs to be is far from clear. More work needs to be done in environmental ethics in order for us to reach sound normative judgments about tough cases like the kinds of trade-offs that an engineered climate is sure to engender.

§5 SUMMARY

After these short suggestions of avenues down which we can pursue future research, a quick recap of the book is in order. I note in Chapter 1 that the purpose of this book was to examine some of the ethical and political complexities engendered by the prospect of geoengineering. The hope was that by analyzing and scrutinizing these complexities we could gain some ground in arriving at appropriate normative judgments about the technology. In Chapter 2, I argue that, despite worries about moral hazards and slippery slopes, research into geoengineering should go forward. I argue that the supposed hazard was empirically uncertain, and that, even if such a hazard were to exist, we would want to know whether the hazard isn't offset by the potential benefits that research could yield. The slippery slope argument, remember, had two premises: that research would lead to deployment (the empirical premise); and that we had decisive moral reasons to avoid deployment (the normative premise). Both premises, I put forward, require more justification. We would both need some kind of evidence to support the empirical claim that research would lead to deployment, and we would also need to know more about SAI to know whether we have decisive moral reasons to avoid its deployment. Ironically, in order to be in a better position to appraise each premise, it is exactly more research that is needed. The conclusion from Chapter 2 was that these arguments shouldn't count as a sufficient ground against research.

But would we ever want to actually deploy SAI? In Chapter 3 I argue that the ideas of respect for nature, precaution, and playing God could not ground a moratorium on future deployment of the technology. Engineering the climate seems like the kind of thing that would count as "disrespecting nature." But Jamieson's approach seems right: we should perhaps in general be more modest when it comes to our domination of nature, but sometimes we are justified in manipulating nature for our purposes. But even if one buys this anthropocentric reasoning, one might think that the precautionary principle should guide our thinking about climate engineering. I argue that when the precautionary principle is vaguely formulated, it is attractive but fails to help guide our action. But when it is made more specific—for instance, when it takes the form of minimax reasoning or a Hartzel-Nichols-type of catastrophic precautionary principle—the conditions that make the principle a reasonable guide for action don't hold. And, finally, worries about the prospect of climate engineering creating novel distributions of winners and losers was explored while looking at the playing God argument. I cast doubt upon the sharp distinction between doing and allowing at the institutional level, and explored how the doctrine of double effect could serve as a possible justification for a decision to deploy, even if this would be worse for some. The point of Chapter 3 was not to show that there are no morally troubling aspects of deploying climate engineering technology, rather, the point was to show that, despite some worries that have been expressed, it would at least be possible for deployment to be morally permissible. The fact that geoengineering may be able to deliver significant benefits, curtail the devastating environmental effects of climate change, and improve the lot of many—especially those most vulnerable to climate change—gives us reason to continue research and think about justifiable deployment scenarios.

But we shouldn't be naïve and assume that just because the technology has the potential to lessen the impact of climate change, alleviate suffering, and further the cause of justice, that it will be researched, developed, and deployed in this way. Indeed, there is always the chance that the technology will be used to further the interests of the few at the expense of the many. For this reason, legitimate regulation of the technology is a must. In Chapter 4 I outline a broad concept of institutional legitimacy and propose appropriate normative criteria to guide our judgments about legitimate geoengineering governance. For an institution overseeing climate engineering to be justifiably labeled legitimate, it should deliver a comparative benefit in a transparent way while also being accountable to those whose action it is attempting to coordinate. Most importantly, the institution must do so in a way that sufficiently satisfies the demands of substantive and procedural justice.

Chapter 5 is devoted to exploring the normative criterion of substantive justice. The fact that climate engineering could create a world with, on average, fewer climate perturbations in the face of anthropogenic climate change than one without such an intervention is interesting. But averages can hide things that are of moral importance. This is why Chapter 5 focuses on the *distribution* of potential benefits and burdens of SAI. I argue that substantive justice requires the benefits and burdens of the pursuit or abandonment of the technology to be skewed heavily in favor of the least well-off members of the global community. This is because they bear the least causal responsibility, the least beneficiary responsibility, and are most vulnerable to the threats of climate change. These three facts point toward the conclusion that it would be unjust to foist significant burdens upon them, even if doing so led to a world that was, on average, better off.

Finally, in the previous chapter we saw that the substantive outcome is not our only concern—we also care about who gets to make decisions regarding geoengineering and how those decisions are made. That is, along with concerns about the distribution of substantive benefits and burdens, we should also pay close attention to the distribution of decision-making power. What we want is a decision-making process that is governed by terms that no one could reasonably reject. More specifically, we want a decision-making process that provides fair terms of inclusion and fair terms of participation. I argue that decision-making power over intentionally manipulating the climate system should include everyone affected by the decision, and should apportion that power according to the extent that each is affected.

§6 FINAL REMARKS

What should be clear by now is that even just the prospect of intentionally manipulating the planetary environment gives rise to some incredibly pernicious and intractable normative difficulties. We would be in a much better situation if previous generations had recognized the climate change problem for what it would become. Still, with any luck, geoengineering will never need to be deployed. Perhaps some of our fundamental assumptions about the climate system are mistaken; perhaps climate change will not carry with it the grave harms to human health and the natural environment that our models currently predict. Or maybe the global community will change course and start seriously cutting emissions and investing in adaptation and negative emission technologies. Such a change would be welcomed with open arms. Unfortunately, even at the writing of this book, such a change still looks like a pious wish.[32]

In the event that it does make sense to include geoengineering as part of our response to climate change, we will have many decisions to make. Some of the decisions will require technical knowledge about the natural world. But even with all the answers to the technical problems of geoengineering, we would still be in the dark with respect to the question of *What should we do?* An answer to this question requires a careful analysis of the ethical and political complexities raised by the prospect of engineering the climate. And, especially given our less-than-perfect response to climate change thus far, it is of the utmost importance that we get the ethics and politics of geoengineering right. Hopefully, this book has made some modest contribution to doing just that.

NOTES

1. Gavin Schmidt, "30 Years after Hansen's Testimony," *Real Climate* (blog), accessed July 1, 2018, http://www.realclimate.org/index.php/archives/2018/06/30-years-after-hansens-testimony/.

2. National Oceanic and Atmospheric Association, "NOAA/ESRL Global Monitoring Division – The NOAA Annual Greenhouse Gas Index (AGGI)," accessed July 20, 2018, https://www.esrl.noaa.gov/gmd/aggi/aggi.html.

3. Many of the concerns I present in the following paragraphs are laid out in more detail in a paper I coauthored with Darrel Moellendorf. See: Daniel Edward Callies and Darrel Moellendorf, "A Moral Assessment of the Available Alternatives to Weak Mitigation Ambition."

4. National Oceanic and Atmospheric Association, "NOAA/ESRL Global Monitoring Division – The NOAA Annual Greenhouse Gas Index (AGGI)."

5. Intergovernmental Panel on Climate Change, *IPCC, 2013: Summary for Policymakers. In: Climate Change 2013: The Physical Science Basis. Contribution of Working Group I to the Fifth Assessment Report of the Intergovernmental Panel on Climate Change* (Cambridge: Cambridge University Press, 2013), http://www.ipcc.ch/pdf/assessment-report/ar5/wg1/WG1AR5_SPM_FINAL.pdf.

6. Intergovernmental Panel on Climate Change, *IPCC, 2018: Summary for Policymakers. In: Global Warming of 1.5 C, an IPCC Special Report on the Impacts of Global Warming of 1.5 above Pre-Industrial Levels and Related Global Greenhouse Gas Emission Pathways, in the Context of Strengthening the Global Response to the Threat of Climate Change, Sustainable Development, and Efforts to Eradicate Poverty* (Cambridge: Cambridge University Press, 2018), http://www.ipcc.ch/pdf/assessment-report/ar5/wg3/ipcc_wg3_ar5_summary-for-policymakers.pdf.

7. Dominic Lenzi et al., "Don't Deploy Negative Emissions Technologies without Ethical Analysis," *Nature* 561, no. 7723 (September 2018): 303–5, https://doi.org/10.1038/d41586-018-06695-5.

8. *Carbon Brief*, "Analysis: Global CO2 Emissions Set to Rise 2% in 2017 after Three-Year 'Plateau,'" *Carbon Brief*, November 13, 2017, https://www.carbonbrief.org/analysis-global-co2-emissions-set-to-rise-2-percent-in-2017-following-three-year-plateau.

9. K. Anderson and A. Bows, "Beyond 'dangerous' Climate Change: Emission Scenarios for a New World," *Philosophical Transactions of the Royal Society A: Mathematical, Physical and Engineering Sciences* 369, no. 1934 (January 13, 2011): 20–44, https://doi.org/10.1098/rsta.2010.0290.

10. Henry Shue, "Climate Surprises: Risk Transfers, Negative Emissions, and the Pivotal Generation," *SSRN Electronic Journal*, 2018, https://doi.org/10.2139/ssrn.3165064.

11. John Rawls, *A Theory of Justice* (Cambridge, MA: Harvard University Press, 1999), 360.

12. Stephen Mark Gardiner, *A Perfect Moral Storm: The Ethical Tragedy of Climate Change*, Environmental Ethics and Science Policy Series (New York: Oxford University Press, 2011).

13. Wilfred Beckerman, "The Impossibility of a Theory of Intergenerational Justice," in *Handbook of Intergenerational Justice*, ed. Jörg Tremmel (Cheltenham, UK: Edward Elgar, 2006).

14. John Stuart Mill, *On Liberty and Other Essays*, Oxford World's Classics (Oxford: Oxford University Press, 2008), 176–203.

15. There are, of course, many other criticisms one could make. For instance, even the currency of utility or happiness strikes many people as an implausible currency for justice to focus on. But I set this and other criticisms aside and focus on the idea of distribution here. And I should highlight that utilitarians have responses to the charge that their theory "disregards distributions," though these responses are not always convincing.

16. Rawls, *A Theory of Justice*, 3.

17. John Rawls, *Political Liberalism* (New York: Columbia University Press, 2005), 274.

18. Peter Singer, *The Expanding Circle: Ethics, Evolution, and Moral Progress*, 1st Princeton University Press pbk. ed (Princeton, NJ: Princeton University Press, 2011).

19. The following is similar to the analysis offered in Toby Svoboda, "The Ethics of Geoengineering: Moral Considerability and the Convergence Hypothesis: The Ethics of Geoengineering," *Journal of Applied Philosophy* 29, no. 3 (August 2012): 243–56, https://doi.org/10.1111/j.1468-5930.2012.00568.x.

20. Peter Singer, "All Animals Are Equal," in *Environmental Ethics: What Really Matters, What Really Works*, ed. David Schmidtz and Elizabeth Willott, 2nd ed. (New York: Oxford University Press, 2012), 49–60.

21. Jeremy Bentham, *An Introduction to the Principles of Morals and Legislation* (New York: Barnes & Noble, 2008).

22. Singer, "All Animals Are Equal," 53.

23. For such a discourse theortic view about our obligations regarding nature, see Benjamin Hale, *The Wild and the Wicked: On Nature and Human Nature* (Cambridge, MA: MIT Press, 2016).

24. Aldo Leopold, *A Sand County Almanac and Sketches Here and There* (New York: Oxford University Press, 1972).

25. Christopher Stone, "Should Trees Have Standing?" in *Environmental Ethics: What Really Matters, What Really Works*, ed. David Schmidtz and Elizabeth Willott, 2nd ed. (New York: Oxford University Press, 2012), 86.

26. Christopher H. Trisos et al., "Potentially Dangerous Consequences for Biodiversity of Solar Geoengineering Implementation and Termination," *Nature Ecology & Evolution* 2, no. 3 (March 2018): 475–82, https://doi.org/10.1038/s41559-017-0431-0.

27. I should also point out that the assumptions built into the model for Trisos et al. are the least optimistic assumptions one could incorporate for an assessment. The model abruptly starts a large-scale geoengineering intervention and that abruptly halts it. This is probably one of the worst-case scenarios one could come up with for SAI.

28. Bryan G. Norton, *Toward Unity among Environmentalists* (New York: Oxford University Press, 1994).

29. Svoboda, "The Ethics of Geoengineering."

30. J. Baird Callicott, "Beyond the Land Ethic," in *Reflecting on Nature: Readings in Environmental Ethics and Philosophy*, ed. Lori Gruen, Dale Jamieson, and Christopher Schlottmann, 2nd ed. (New York: Oxford University Press, 2013), 86.

31. Callicott, 86.

32. Paul J. Crutzen, "Albedo Enhancement by Stratospheric Sulfur Injections: A Contribution to Resolve a Policy Dilemma?," *Climatic Change* 77, no. 3–4 (September 1, 2006): 211–20, https://doi.org/10.1007/s10584-006-9101-y.

Bibliography

Adams, N. P. "Institutional Legitimacy." *Journal of Political Philosophy* 26, no. 1 (March 2018): 84–102. https://doi.org/10.1111/jopp.12122.
Anderson, K., and A. Bows. "Beyond 'dangerous' Climate Change: Emission Scenarios for a New World." *Philosophical Transactions of the Royal Society A: Mathematical, Physical and Engineering Sciences* 369, no. 1934 (January 13, 2011): 20–44. https://doi.org/10.1098/rsta.2010.0290.
Aquinas, Saint Thomas. *Summa Theologica*, 1265.
Aristotle. *Nicomachean Ethics*, n.d.
Arneson, Richard. "Democracy Is Not Intrinsically Just." In *Philosophy, Politics and Society*, edited by Peter Laslett and James Fishkin. 5. New Haven, CT: Yale University Press, 1979.
———. "Democratic Rights at the National and Workplace Levels." In *The Idea of Democracy*, edited by David Copp, Jean Hampton, and John E. Roemer. New York: Cambridge University Press, 1995.
Arrhenius, Gustaf. "Democracy for the 21st Century." *Sociology Looks at the Twenty-First Century*, 2013.
———. "The Boundary Problem in Democratic Theory." *Democracy Unbound: Basic Explorations I*, 2005, 14–29.
Ashford et al., Nicholas. "The Wingspread Statement on the Precautionary Principle," January 1998.
Bannister, Frank, and Regina Connolly. "The Trouble with Transparency: A Critical Review of Openness in e-Government." *Policy & Internet* 3, no. 1 (February 1, 2011): 1–30. https://doi.org/10.2202/1944-2866.1076.
Barrett, Scott. *Environment and Statecraft*. Oxford: Oxford University Press, 2005.
———. "The Incredible Economics of Geoengineering." *Environmental and Resource Economics* 39, no. 1 (January 1, 2008): 45–54. https://doi.org/10.1007/s10640-007-9174-8.
Beckerman, Wilfred. "The Impossibility of a Theory of Intergenerational Justice." In *Handbook of Intergenerational Justice*, edited by Jörg Tremmel. Cheltenham, UK: Edward Elgar, 2006.
Beerbohm, Eric. *In Our Name: The Ethics of Democracy*. Princeton, NJ: Princeton University Press, 2015.
Beitz, Charles R. *Political Equality: An Essay in Democratic Theory*. Princeton, NJ: Princeton University Press, 1989.
Bentham, Jeremy. *An Introduction to the Principles of Morals and Legislation*. New York: Barnes & Noble, 2008.

Berkey, Brian. "State Action, State Policy, and the Doing/Allowing Distinction." *Ethics, Policy & Environment* 17, no. 2 (May 4, 2014): 147–49. https://doi.org/10.1080/21550085.2014.926074.

Blake, Michael. "Distributive Justice, State Coercion, and Autonomy." *Philosophy & Public Affairs* 30, no. 3 (July 1, 2001): 257–96. https://doi.org/10.1111/j.1088-4963.2001.00257.x.

Bodansky, Daniel. "May We Engineer the Climate?" *Climatic Change* 33, no. 3 (July 1, 1996): 309–21. https://doi.org/10.1007/BF00142579.

———. "The Who, What, and Wherefore of Geoengineering Governance." *Climatic Change* 121, no. 3 (December 2013): 539–51. https://doi.org/10.1007/s10584-013-0759-7.

Brighouse, Harry. *Justice*. Key Concepts. Cambridge: Polity, 2004.

Brighouse, Harry, and Marc Fleurbaey. "Democracy and Proportionality." *Journal of Political Philosophy* 18, no. 2 (June 2010): 137–55. https://doi.org/10.1111/j.1467-9760.2008.00316.x.

Buchanan, Allen. "Institutional Legitimacy." In *Oxford Studies in Political Philosophy*, edited by David Sobel, Peter Vallentyne, and Steven Wall, 4:53–78. Oxford: Oxford University Press, 2018.

———. *The Heart of Human Rights*. Oxford: Oxford University Press, 2013. https://doi.org/10.1093/acprof:oso/9780199325382.001.0001.

Buchanan, Allen E. *Justice, Legitimacy, and Self-Determination: Moral Foundations for International Law*. Oxford Political Theory. Oxford: Oxford University Press, 2004.

Buchanan, Allen, and Robert O. Keohane. "The Legitimacy of Global Governance Institutions." *Ethics & International Affairs* 20, no. 4 (2006): 405–37.

Bullard, Robert. *The Quest for Environmental Justice: Human Rights and the Politics of Pollution*. San Francisco: Sierra Club Books, 2005.

Bunzl, Martin. "An Ethical Assessment of Geoengineering." *Bulletin of the Atomic Scientists*, June 2008, 18.

Burg, Wibren van der. "The Slippery Slope Argument." *Ethics* 102, no. 1 (October 1991): 42–65. https://doi.org/10.1086/293369.

Burgess, J. A. "The Great Slippery-Slope Argument." *Journal of Medical Ethics* 19, no. 3 (September 1, 1993): 169–74. https://doi.org/10.1136/jme.19.3.169.

Burns, Elizabeth T., Jane A. Flegal, David W. Keith, Aseem Mahajan, Dustin Tingley, and Gernot Wagner. "What Do People Think When They Think about Solar Geoengineering? A Review of Empirical Social Science Literature, and Prospects for Future Research." *Earth's Future* 4, no. 11 (November 1, 2016): 536–42. https://doi.org/10.1002/2016EF000461.

Callicott, J. Baird. "Beyond the Land Ethic." In *Reflecting on Nature: Readings in Environmental Ethics and Philosophy*, edited by Lori Gruen, Dale Jamieson, and Christopher Schlottmann, 2nd edition. New York: Oxford University Press, 2013.

Callies, Daniel Edward. "Institutional Legitimacy and Geoengineering Governance." *Ethics, Policy & Environment*, 2019. DOI: 10.1080/21550085.2018.1562523.

———. "The Slippery Slope Argument against Geoengineering Research." *Journal of Applied Philosophy*, 2018. DOI: 10.1111/japp.12345.

———. "Paris Agreement Bigger than Any One Man." *Agenda for International Development*, June 2017. http://www.a-id.org/en/pubblicazioni/.

Callies, Daniel Edward, and Darrel Moellendorf. "A Moral Assessment of the Available Alternatives to Weak Mitigation Ambition." *Working Paper*, n.d.

Caney, Simon. "Cosmopolitan Justice, Responsibility, and Global Climate Change." *Leiden Journal of International Law* 18, no. 04 (January 9, 2006): 747. https://doi.org/10.1017/S0922156505002992.

———. *Justice beyond Borders: A Global Political Theory*. Oxford: Oxford University Press, 2006.

Carbon Brief. "Analysis: Global CO2 Emissions Set to Rise 2% in 2017 after Three-Year 'Plateau.'" *Carbon Brief*, November 13, 2017. https://www.carbonbrief.org/analysis-global-co2-emissions-set-to-rise-2-percent-in-2017-following-three-year-plateau.

Cavanaugh, T. A. *Double-Effect Reasoning: Doing Good and Avoiding Evil*. Oxford: Oxford University Press, 2006.

Christiano, Thomas. *The Constitution of Equality: Democratic Authority and Its Limits*. Oxford: Oxford University Press, 2008.
Cicerone, Ralph J. "Geoengineering: Encouraging Research and Overseeing Implementation." *Climatic Change* 77, no. 3–4 (September 1, 2006): 221–26. https://doi.org/10.1007/s10584-006-9102-x.
Cooper, Robert G. "A Process Model for Industrial New Product Development." *IEEE Transactions on Engineering Management* EM-30, no. 1 (1983): 2–11. https://doi.org/10.1109/TEM.1983.6448637.
———. "Stage-Gate Systems: A New Tool for Managing New Products." *Business Horizons* 33, no. 3 (May 1990): 44–54. https://doi.org/10.1016/0007-6813(90)90040-I.
Copp, David, Jean Hampton, and John E. Roemer, eds. *The Idea of Democracy*. 1st pbk. ed. Cambridge: Cambridge University Press, 1995.
Crutzen, Paul J. "Albedo Enhancement by Stratospheric Sulfur Injections: A Contribution to Resolve a Policy Dilemma?" *Climatic Change* 77, no. 3–4 (September 1, 2006): 211–20. https://doi.org/10.1007/s10584-006-9101-y.
Cutter, Susan. *Hazards, Vulnerability, and Environmental Justice*. New York: Earthscan, 2006.
Dahl, Robert A. *After the Revolution; Authority in a Good Society*. New Haven, CT: Yale University Press, 1970.
———. "Procedural Democracy." In *Philosophy, Politics and Society*, edited by Peter Laslett and James Fishkin. 5. New Haven, CT: Yale University Press, 1979.
Daniels, Norman. "Reflective Equilibrium." In *The Stanford Encyclopedia of Philosophy*, edited by Edward N. Zalta, Winter 2016. Metaphysics Research Lab, Stanford University, 2016. https://plato.stanford.edu/archives/win2016/entries/reflective-equilibrium/.
———. "Wide Reflective Equilibrium and Theory Acceptance in Ethics." *The Journal of Philosophy* 76, no. 5 (1979): 256–82. https://doi.org/10.2307/2025881.
Dembe, Allard E., and Leslie I. Boden. "Moral Hazard: A Question of Morality?" *New Solutions: A Journal of Environmental and Occupational Health Policy* 10, no. 3 (November 2000): 257–79. https://doi.org/10.2190/1GU8-EQN8-02J6-2RXK.
Douglas, Thomas. "Intertemporal Disagreement and Empirical Slippery Slope Arguments." *Utilitas* 22, no. 2 (June 2010): 184–97. https://doi.org/10.1017/S0953820810000087.
Dworkin, Ronald. *Justice for Hedgehogs*. Cambridge, MA: Belknap Press of Harvard University Press, 2011.
———. *Taking Rights Seriously*. London: Bloomsbury, 2013.
Edmonds, David. *Would You Kill the Fat Man?: The Trolley Problem and What Your Answer Tells Us about Right and Wrong*. Princeton, NJ: Princeton University Press, 2014.
Elliott, Kevin. "An Ethics of Expertise Based on Informed Consent." *Science and Engineering Ethics* 12, no. 4 (March, 2006): 637–61. https://doi.org/10.1007/S11948-006-0062-3.
Environmental Protection Agency. "Health and Environmental Effects of Ozone Layer Depletion." Reports and Assessments, July 17, 2015. https://www.epa.gov/ozone-layer-protection/health-and-environmental-effects-ozone-layer-depletion.
———, n.d. https://www.epa.gov/superfund/public-comment-process.
Etzioni, Amitai. "Is Transparency the Best Disinfectant?" *Journal of Political Philosophy* 18, no. 4 (December 1, 2010): 389–404. https://doi.org/10.1111/j.1467-9760.2010.00366.x.
Food and Agriculture Organization of the United Nations. *The State of the World Fisheries and Aquaculture 2014: Opportunities and Challenges*. Rome: Food and Agriculture Organization of the United Nations, 2014.
Foot, Philippa. "The Problem of Abortion and the Doctrine of Double Effect." In *Virtues and Vices*. Oxford: Oxford University Press, 2002.
———. *Virtues and Vices*. Oxford: Oxford University Press, 2002.
Forst, Rainer. *The Right to Justification: Elements of a Constructivist Theory of Justice*. Translated by Jeffrey Flynn. New York: Columbia University Press, 2012.
Gardiner, Stephen. *A Perfect Moral Storm: The Ethical Tragedy of Climate Change*. New York: Oxford University Press, 2011.
———. "Is Arming the Future with Geoengineering Really the Lesser Evil?" In *Climate Ethics*, edited by Stephen Gardiner, Simon Caney, Dale Jamieson, and Henry Shue, 284–312. Oxford: Oxford University Press, 2010.

Gardiner, Stephen M. "A Core Precautionary Principle*." *Journal of Political Philosophy* 14, no. 1 (March 1, 2006): 33–60. https://doi.org/10.1111/j.1467-9760.2006.00237.x.

———. "A Perfect Moral Storm: Climate Change, Intergenerational Ethics and the Problem of Moral Corruption." *Environmental Values* 15, no. 3 (August 1, 2006): 397–413. https://doi.org/10.3197/096327106778226293.

———. "Geoengineering: Ethical Questions for Deliberate Climate Manipulators." In *The Oxford Handbook of Environmental Ethics*, edited by Stephen M. Gardiner and Allen Thompson. Oxford: Oxford University Press, 2016.

———. "The Desperation Argument for Geoengineering." *Political Science & Politics* 46, no. 1 (January 2013): 28–33. https://doi.org/10.1017/S1049096512001424.

———. "Why Geoengineering Is Not a 'Global Public Good,' and Why It Is Ethically Misleading to Frame It as One." *Climatic Change* 121, no. 3 (December 1, 2013): 513–25. https://doi.org/10.1007/s10584-013-0764-x.

Gardiner, Stephen M., and Allen Thompson, eds. *The Oxford Handbook of Environmental Ethics*. Oxford: Oxford University Press, 2016.

Gardiner, Stephen Mark. *A Perfect Moral Storm: The Ethical Tragedy of Climate Change*. Environmental Ethics and Science Policy Series. New York: Oxford University Press, 2011.

———. *Debating Climate Ethics*. Debating Ethics. New York: Oxford University Press, 2016.

Gardiner, Stephen Mark, Simon Caney, Dale Jamieson, and Henry Shue, eds. *Climate Ethics: Essential Readings*. Oxford: Oxford University Press, 2010.

Gardiner, Stephen M. "Ethics and Global Climate Change." *Ethics* 114, no. 3 (April 2004): 555–600. https://doi.org/10.1086/382247.

Gasser, T., C. Guivarch, K. Tachiiri, C. D. Jones, and P. Ciais. "Negative Emissions Physically Needed to Keep Global Warming below 2 °C." *Nature Communications* 6 (August 3, 2015): 7958. https://doi.org/10.1038/ncomms8958.

Goodin, Robert E. "Enfranchising All Affected Interests, and Its Alternatives." *Philosophy & Public Affairs* 35, no. 1 (2007): 40–68.

———. "Enfranchising the Earth, and Its Alternatives." *Political Studies* 44, no. 5 (1996): 835–49.

Griffin, Christopher G. "Democracy as a Non–Instrumentally Just Procedure." *Journal of Political Philosophy* 11, no. 1 (2003): 111–21.

Gruen, Lori, Dale Jamieson, and Christopher Schlottmann, eds. *Reflecting on Nature: Readings in Environmental Ethics and Philosophy*. 2nd edition. New York: Oxford University Press, 2013.

Hale, Ben. "The World That Would Have Been: Moral Hazard Arguments Against Geoengineering." In *Engineering the Climate: The Ethics of Solar Radiation Management*, edited by Christopher Preston, 113–31. Lanham, MD: Lexington Books, 2012.

Hale, Benjamin. *The Wild and the Wicked: On Nature and Human Nature*. Cambridge, MA: MIT Press, 2016.

Hamilton, Clive. *Earthmasters: The Dawn of the Age of Climate Engineering*. New Haven, CT: Yale University Press, 2014.

———. "No, We Should Not Just 'at Least Do the Research.'" *Nature* 496 (April 2013): 139.

Hampshire, Stuart. "Liberalism: The New Twist." *New York Review of Books*, August 12, 1993.

Hardin, Garrett. "The Tragedy of the Commons." *Science* 162 (December 1968): 1243–48.

Hart, H. L. A. *The Concept of Law*. 2nd edition. Oxford: Oxford University Press, 1998.

Hartzell-Nichols, Lauren. "Precaution and Solar Radiation Management." *Ethics, Policy & Environment* 15, no. 2 (June 2012): 158–71. https://doi.org/10.1080/21550085.2012.685561.

Haywood, Jim M., Andy Jones, Nicolas Bellouin, and David Stephenson. "Asymmetric Forcing from Stratospheric Aerosols Impacts Sahelian Rainfall." *Nature Climate Change* 3, no. 7 (March 31, 2013): 660–65. https://doi.org/10.1038/nclimate1857.

Heyd, Thomas, ed. *Recognizing the Autonomy of Nature: Theory and Practice*. New York: Columbia University Press, 2005.

Heyward, Clare. "Benefiting from Climate Geoengineering and Corresponding Remedial Duties: The Case of Unforeseeable Harms." *Journal of Applied Philosophy* 31, no. 4 (November 1, 2014): 405–19. https://doi.org/10.1111/japp.12075.

———. "Can the All-Affected Principle Include Future Persons? Green Deliberative Democracy and the Non-Identity Problem." *Environmental Politics* 17, no. 4 (August 2008): 625–43. https://doi.org/10.1080/09644010802193591.

———. "Is There Anything New under the Sun? Exceptionalism, Novelty, and Debating Geoengineering Governance." In *The Ethics of Climate Governance*, edited by Aaron Maltais and Catriona McKinnon. Lanham, MD: Rowman & Littlefield Publishers, 2015.

———. "Situating and Abandoning Geoengineering: A Typology of Five Responses to Dangerous Climate Change." *PS: Political Science & Politics* 46, no. 1 (January 2013): 23–27. https://doi.org/10.1017/S1049096512001436.

Heyward, Clare, and Dominic Roser, eds. *Climate Justice in a Non-Ideal World*. Oxford: Oxford University Press, 2016. doi.org/10.1093/acprof:oso/9780198744047.001.0001.

Honoré, Tony. *Responsibility and Fault*. Oxford: Hart Publishing, 2002.

Horton, Joshua. "Geoengineering and the Myth of Unilateralism: Pressures and Prospects for International Cooperation." *Stanford Journal of Law, Science & Policy* 4 (2011): 56–69.

Horton, Joshua B., and David Keith. "Solar Geoengineering and Obligations to the Global Poor." In *Climate Justice and Geoengineering: Ethics and Policy in the Atmospheric Anthropocene*, edited by Christopher J. Preston, 79–92. London: Rowman & Littlefield International, Ltd, 2016.

Horton, Joshua B., Jesse L. Reynolds, Holly Jean Buck, Daniel Edward Callies, Stefan Schaefer, David W. Keith, and Steve Rayner. "Solar Geoengineering and Democracy." *Global Environmental Politics* 18, no. 3 (August 2018).

Hulme, Mike. *Can Science Fix Climate Change? A Case against Climate Engineering*. New Human Frontiers Series. Cambridge: Polity Press, 2014.

Intergovernmental Panel on Climate Change. *Climate Change 2001: Synthesis Report*. Edited by R. T. Watson, Daniel L. Albritton, Intergovernmental Panel on Climate Change, and Intergovernmental Panel on Climate Change. Cambridge: Cambridge University Press, 2001.

———. *Climate Change 2014: Synthesis Report*. Edited by R. K. Pachauri, Leo Mayer, and Intergovernmental Panel on Climate Change. Geneva, Switzerland: Intergovernmental Panel on Climate Change, 2015.

———. *Climate Change: The IPCC Scientific Assessment*. Edited by John Theodore Houghton, G. J. Jenkins, and J. J. Ephraums. Cambridge: Cambridge University Press, 1990.

———. *IPCC, 2013: Summary for Policymakers. In: Climate Change 2013: The Physical Science Basis. Contribution of Working Group I to the Fifth Assessment Report of the Intergovernmental Panel on Climate Change*. Cambridge: Cambridge University Press, 2013. http://www.ipcc.ch/pdf/assessment-report/ar5/wg1/WG1AR5_SPM_FINAL.pdf.

———. *IPCC, 2014: Summary for Policymakers. In: Climate Change 2014: Impacts, Adaptation, and Vulnerability. Part A: Global and Sectoral Aspects. Contribution of Working Group II to the Fifth Assessment Report of the Intergovernmental Panel on Climate Change*. Cambridge: Cambridge University Press, 2014. http://www.ipcc.ch/pdf/assessment-report/ar5/wg2/ar5_wgII_spm_en.pdf.

———. *IPCC, 2014: Summary for Policymakers. In: Climate Change 2014: Mitigation of Climate Change. Contribution of Working Group III to the Fifth Assessment Report of the Intergovernmental Panel on Climate Change*. Cambridge: Cambridge University Press, 2014. http://www.ipcc.ch/pdf/assessment-report/ar5/wg3/ipcc_wg3_ar5_summary-for-policymakers.pdf.

———. *IPCC, 2018: Summary for Policymakers. In: Global Warming of 1.5 C, an IPCC Special Report on the Impacts of Global Warming of 1.5 above Pre-Industrial Levels and Related Global Greenhouse Gas Emission Pathways, in the Context of Strengthening the Global Response to the Threat of Climate Change, Sustainable Development, and Efforts to Eradicate Poverty*. Cambridge: Cambridge University Press, 2018. http://www.ipcc.ch/pdf/assessment-report/ar5/wg3/ipcc_wg3_ar5_summary-for-policymakers.pdf.

International Atomic Energy Association. "Country Nuclear Power Profiles." Accessed February 2, 2018. https://cnpp.iaea.org/countryprofiles/Germany/Germany.htm.

International Monetary Fund. "World Economic Outlook Database 2016." Accessed August 18, 2017. http://www.imf.org/external/pubs/ft/weo/2016/01/weodata/weorept.aspx.

Irvine, Peter J., Andy Ridgwell, and Daniel J. Lunt. "Assessing the Regional Disparities in Geoengineering Impacts." *Geophysical Research Letters* 37, no. 18 (September 2010): 1–6. https://doi.org/10.1029/2010GL044447.

Jamieson, Dale. "Climate Change, Responsibility, and Justice." *Science and Engineering Ethics* 16, no. 3 (September 2010): 431–45. https://doi.org/10.1007/s11948-009-9174-x.

———. "Ethics and Intentional Climate Change." *Climatic Change* 33, no. 3 (July 1, 1996): 323–36. https://doi.org/10.1007/BF00142580.

———. *Reason in a Dark Time: Why the Struggle Against Climate Change Failed—and What It Means for Our Future*. Oxford: Oxford University Press, 2014.

Janssen, M. A., and Michel den Elzen. "Allocating C02-Emissions by Using Equity Rules and Optimization." *National Institute of Public Health and Environmental Protection – Bilthoven, The Netherlands*, 1992.

Jefferson, Anneli. "Slippery Slope Arguments." *Philosophy Compass* 9, no. 10 (October 2014): 672–80. https://doi.org/10.1111/phc3.12161.

Joerges, Bernward. "Do Politics Have Artefacts?" *Social Studies of Science* 29, no. 3 (June 1999): 411–31. https://doi.org/10.1177/030631299029003004.

Johnson, Jim. "Mixing Humans with Non-Humans: The Sociology of a Door-Closer." *Social Problems* 35 (1988): 298–310.

Kahan, Dan M., Hank C. Jenkins-Smith, Tor Tarantola, Carol L Silva, and Donald Braman. "Geoengineering and the Science Communication Environment: A Cross-Cultural Experiment." *Annals of the American Academy of Political and Social Science*, 2014. https://doi.org/10.2139/ssrn.1981907.

Kamm, Frances M. "Harming Some to Save Others." *Philosophical Studies* 57, no. 3 (1989): 227–260.

Kass, Leon, and James Q. Wilson. *The Ethics of Human Cloning*. Washington, DC: AEI Press, 1998.

Kates, Robert W. "Cautionary Tales: Adaptation and the Global Poor." *Climatic Change* 45, no. 1 (April 1, 2000): 5–17. https://doi.org/10.1023/A:1005672413880.

Katz, Eric. *Nature as Subject: Human Obligation and Natural Community*. Studies in Social, Political, and Legal Philosophy. Lanham, MD: Rowman & Littlefield, 1997.

Keith, David. *A Critical Look at Geoengineering against Climate Change*. Accessed August 18, 2017. https://www.ted.com/talks/david_keith_s_surprising_ideas_on_climate_change.

Keith, David W. *A Case for Climate Engineering*. Boston Review Books. Cambridge, MA: The MIT Press, 2013.

———. "Geoengineering the Climate: History and Prospect." *Annual Review of Energy and the Environment* 25, no. 1 (2000): 245–84. https://doi.org/10.1146/annurev.energy.25.1.245.

Keith, David W., Geoffrey Holmes, David St. Angelo, and Kenton Heidel. "A Process for Capturing CO_2 from the Atmosphere." *Joule*, June 2018. https://doi.org/10.1016/j.joule.2018.05.006.

Keith, David W., and Douglas G. MacMartin. "A Temporary, Moderate and Responsive Scenario for Solar Geoengineering." *Nature Climate Change* 5, no. 3 (February 16, 2015): 201–6. https://doi.org/10.1038/nclimate2493.

Keith, David W., Edward Parson, and M. Granger Morgan. "Research on Global Sunblock Needed Now." *Nature* 463, no. 7280 (January 28, 2010): 426–27. https://doi.org/10.1038/463426a.

Keith, David W., Debra K. Weisenstein, John A. Dykema, and Frank N. Keutsch. "Stratospheric Solar Geoengineering without Ozone Loss." *Proceedings of the National Academy of Sciences* 113, no. 52 (December 27, 2016): 14910–14. https://doi.org/10.1073/pnas.1615572113.

Kellogg, W. W., and S. H. Schneider. "Climate Stabilization: For Better or for Worse?" *Science* 186, no. 4170 (December 27, 1974): 1163–72. https://doi.org/10.1126/science.186.4170.1163.

Kemp, Luke. "Framework for the Future? Exploring the Possibility of Majority Voting in the Climate Negotiations." *International Environmental Agreements: Politics, Law and Economics* 16, no. 5 (October 2016): 757–79. https://doi.org/10.1007/s10784-015-9294-5.

Kola, Ismail, and John Landis. "Can the Pharmaceutical Industry Reduce Attrition Rates?" *Nature Reviews Drug Discovery* 3, no. 8 (August 2004): 711–16. https://doi.org/10.1038/nrd1470.

Korsgaard, Christine M. "Realism and Constructivism in Twentieth-Century Moral Philosophy." *Journal of Philosophical Research* 28 (2003): 99–122.

LaFrance, Adrienne. "Genetically Modified Mosquitoes: What Could Possibly Go Wrong?" *The Atlantic*, April 26, 2016. https://www.theatlantic.com/technology/archive/2016/04/genetically-modified-mosquitoes-zika/479793/.

Lane, Lee, Ken Caldeira, Robert Chatfield, and Stephanie Langhoff. "Workshop Report on Managing Solar Radiation," April 1, 2007. ntrs.nasa.gov/search.jsp?R=20070031204.

Laslett, Peter, and James Fishkin, eds. *Philosophy, Politics and Society*. Philosophy, Politics and Society : A Collection 5. New Haven, CT: Yale University Press [u.a.], 1979.

Lenzi, Dominic, William F. Lamb, Jérôme Hilaire, Martin Kowarsch, and Jan C. Minx. "Don't Deploy Negative Emissions Technologies without Ethical Analysis." *Nature* 561, no. 7723 (September 2018): 303–5. https://doi.org/10.1038/d41586-018-06695-5.

Leopold, Aldo. *A Sand County Almanac and Sketches Here and There*. New York: Oxford University Press, 1972.

Lin, Albert C. "Geoengineering Governance." *Issues in Legal Scholarship* 8, no. 1 (January 13, 2009). https://doi.org/10.2202/1539-8323.1112.

———. "The Missing Pieces of Geoengineering Research Governance." *Minnesota Law Review* 100, no. 6 (2016): 2509–76.

Lloyd, William Foster. "W. F. Lloyd on the Checks to Population." *Population and Development Review* 6, no. 3 (September 1980): 473–96. https://doi.org/10.2307/1972412.

Lomas, A., J. Leonardi-Bee, and F. Bath-Hextall. "A Systematic Review of Worldwide Incidence of Nonmelanoma Skin Cancer." *British Journal of Dermatology* 166, no. 5 (May 1, 2012): 1069–80. https://doi.org/10.1111/j.1365-2133.2012.10830.x.

Long, Jane. "Task Force on Climate Remediation Research," October 4, 2011. https://bipartisanpolicy.org/library/task-force-climate-remediation-research/.

Macnaghten, Phil, and Bronislaw Szerszynski. "Living the Global Social Experiment: An Analysis of Public Discourse on Solar Radiation Management and Its Implications for Governance." *Global Environmental Change* 23, no. 2 (April 2013): 465–74.

Maffettone, Pietro and Luke Ulaş. "Legitimacy, Metacoordination, and Context-Dependence." *International Theory* 11 (2019): 81–109.

Mangan, Joseph T. "An Historical Analysis of the Principle of Double Effect." *Theological Studies* 10, no. 1 (February 1949): 41–61. https://doi.org/10.1177/004056394901000102.

Manson, Neil A. "Formulating the Precautionary Principle." *Environmental Ethics* 24, no. 3 (2002): 263–274.

Maskin, Eric, and Jean Tirole. "The Politician and the Judge: Accountability in Government." *American Economic Review* 94, no. 4 (August 2004): 1034–54. https://doi.org/10.1257/0002828042002606.

McIntyre, Alison. "Doctrine of Double Effect." In *The Stanford Encyclopedia of Philosophy*, edited by Edward N. Zalta, Winter 2014. Metaphysics Research Lab, Stanford University, 2014. https://plato.stanford.edu/archives/win2014/entries/double-effect/.

McMahan, Jeff. "Killing, Letting Die, and Withdrawing Aid." *Ethics* 103, no. 2 (1993): 250–79. https://doi.org/10.2307/2381522.

Mercer, A. M., D. W. Keith, and J. D. Sharp. "Public Understanding of Solar Radiation Management." *Environmental Research Letters* 6, no. 4 (October 1, 2011): 044006. https://doi.org/10.1088/1748-9326/6/4/044006.

Mikhail, John. "Universal Moral Grammar: Theory, Evidence, and the Future." *Trends in Cognitive Sciences* 11, no. 4 (April 2007): 143–52.

Mill, John Stuart. *On Liberty and Other Essays*. Oxford World's Classics. Oxford: Oxford University Press, 2008.

Miller, David. *National Responsibility and Global Justice*. Oxford Political Theory. Oxford: Oxford University Press, 2012.

Miller, Richard W. "Cosmopolitan Respect and Patriotic Concern." *Philosophy and Public Affairs* 27, no. 3 (July 1998): 202–24. https://doi.org/10.1111/j.1088-4963.1998.tb00068.x.

Moellendorf, Darrel. "A Right to Sustainable Development." *The Monist* 94, no. 3 (2011): 433–52.

———. "Equal Respect and Global Egalitarianism." *Social Theory and Practice* 32, no. 4 (2006): 601–616.

———. "Taking UNFCCC Norms Seriously." In *Climate Justice in a Non-Ideal World*, edited by Clare Heyward and Dominic Roser. Oxford: Oxford University Press, 2016. https://doi.org/10.1093/acprof:oso/9780198744047.001.0001.

———. *The Moral Challenge of Dangerous Climate Change: Values, Poverty, and Policy*. New York: Cambridge University Press, 2014.

Moreno-Cruz, Juan. "Mitigation and the Geoengineering Threat." *Resource and Energy Economics* 41, no. C (2015): 248–63.

Morrow, David. "International Governance of Climate Engineering: A Survey of Reports on Climate Engineering, 2009–2015." *SSRN Electronic Journal*, 2017. https://doi.org/10.2139/ssrn.2982392.

Morrow, David R. "Fairness in Allocating the Global Emissions Budget." *Environmental Values* 26, no. 6 (December 1, 2017): 669–91. doi.org/10.3197/096327117X15046905490335.

———. "Starting a Flood to Stop a Fire? Some Moral Constraints on Solar Radiation Management." *Ethics, Policy & Environment* 17, no. 2 (May 4, 2014): 123–38. https://doi.org/10.1080/21550085.2014.926056.

———. "Why Geoengineering Is a Public Good, Even If It Is Bad." *Climatic Change* 123, no. 2 (March 2014): 95–100. https://doi.org/10.1007/s10584-013-0967-1.

Morrow, David R., Robert Kopp, and Michael Oppenheimer. "Political Legitimacy in Decisions about Experiments in Solar Radiation Management." In *Climate Change Geoengineering: Philosophical Perspectives, Legal Issues, and Governance Frameworks*, edited by William C. G. Burns and Andrew Strauss, 146–67. Cambridge: Cambridge University Press, 2013.

Nagel, Thomas. "The Problem of Global Justice." *Philosophy & Public Affairs* 33, no. 2 (2005): 113–147.

Näsström, Sofia. "The Challenge of the All-Affected Principle." *Political Studies* 59, no. 1 (March 2011): 116–34. https://doi.org/10.1111/j.1467-9248.2010.00845.x.

National Oceanic and Atmospheric Association. "NOAA/ESRL Global Monitoring Division - The NOAA Annual Greenhouse Gas Index (AGGI)." Accessed July 20, 2018. https://www.esrl.noaa.gov/gmd/aggi/aggi.html.

———. "Science—Ozone Basics." Accessed April 29, 2016. http://www.ozonelayer.noaa.gov/science/basics.htm.

National Public Radio. "Chinese Scientists Clone Monkeys Using Method That Created Dolly The Sheep." *All Things Considered*. National Public Radio, January 24, 2018. https://www.npr.org/sections/health-shots/2018/01/24/579925801/chinese-scientists-clone-monkeys-using-method-that-created-dolly-the-sheep.

Natural Environment Research Council. "Experiment Earth? Report on a Public Dialogue on Geoengineering," August 2010.

Neumayer, Eric. "In Defence of Historical Accountability for Greenhouse Gas Emissions." *Ecological Economics* 33, no. 2 (2000): 185–192.

Nicholson, Simon, Sikina Jinnah, and Alexander Gillespie. "Solar Radiation Management: A Proposal for Immediate Polycentric Governance." *Climate Policy* 18, no. 3 (March 16, 2018): 322–34. https://doi.org/10.1080/14693062.2017.1400944.

Nordhaus, William D. *A Question of Balance: Weighing the Options on Global Warming Policies*, 2015.

Nordhaus, William D. *Managing the Global Commons: The Economics of Climate Change*. Cambridge, MA: MIT Press, 1994.

Norton, Bryan G. *Toward Unity among Environmentalists*. New York: Oxford University Press, 1994.

Notre Dame Global Adaptation Index. "Vulnerability Rankings | ND-GAIN Index." Accessed August 18, 2017. http://index.gain.org/ranking/vulnerability.
Nozick, Robert. *Anarchy, State, and Utopia*. Malden, MA: Blackwell, 2012.
Nussbaum, Martha C. "The Costs of Tragedy: Some Moral Limits of Cost-Benefit Analysis." *The Journal of Legal Studies* 29, no. S2 (2000): 1005–36.
Obradovich, Nick. "Climate Change May Speed Democratic Turnover." *Climatic Change* 140, no. 2 (January 2017): 135–47. https://doi.org/10.1007/s10584-016-1833-8.
O'Neill, John, Alan Holland, and Andrew Light. *Environmental Values*. London: Routledge, 2008.
O'Neill, Onora. "Transparency and the Ethics of Communication." In *Transparency: The Key to Better Governance?*, edited by Christopher Hood and David Heald. Proceedings of the British Academy 135. Oxford: Oxford University Press, 2006.
Ott, Konrad. "Might Solar Radiation Management Constitute a Dilemma?" In *Engineering the Climate: The Ethics of Solar Radiation Management*, edited by Christopher J. Preston, 113–31. Lanham, MD: Lexington Books, 2012.
Owen, David. "Constituting the Polity, Constituting the Demos: On the Place of the All Affected Interests Principle in Democratic Theory and in Resolving the Democratic Boundary Problem." *Ethics & Global Politics* 5, no. 3 (January 2012): 129–52. https://doi.org/10.3402/egp.v5i3.18617.
Page, Edward A. "Give It up for Climate Change: A Defence of the Beneficiary Pays Principle." *International Theory* 4, no. 2 (July 2012): 300–330. https://doi.org/10.1017/S175297191200005X.
Parfit, Derek. *On What Matters*. The Berkeley Tanner Lectures. Oxford: Oxford University Press, 2011.
Parfit, Derek. *Reasons and Persons*. Oxford: Clarendon Press, 1984.
Parson, Edward A. "Climate Engineering in Global Climate Governance: Implications for Participation and Linkage." *Transnational Environmental Law* 3, no. 1 (April 2014): 89–110. https://doi.org/10.1017/S2047102513000496.
Pettit, Philip. *Republicanism: A Theory of Freedom and Government*. Oxford: Clarendon Press ; Oxford University Press, 1997. http://site.ebrary.com/id/10273243.
Pielke, Roger, Gwyn Prins, Steve Rayner, and Daniel Sarewitz. "Climate Change 2007: Lifting the Taboo on Adaptation." *Nature* 445, no. 7128 (February 8, 2007): 597–98. https://doi.org/10.1038/445597a.
Posner, Eric A, and David Weisbach. *Climate Change Justice*. Princeton, NJ: Princeton University Press, 2015.
Pounds, J. Alan, Michael P. L. Fogden, and John H. Campbell. "Biological Response to Climate Change on a Tropical Mountain." *Nature* 398, no. 6728 (April 1999): 611–15. https://doi.org/10.1038/19297.
Preston, Christopher J., ed. *Climate Justice and Geoengineering: Ethics and Policy in the Atmospheric Anthropocene*. London: Rowman & Littlefield International, Ltd., 2016.
———, ed. *Engineering the Climate: The Ethics of Solar Radiation Management*. Lanham, MD: Lexington Books, 2012.
———. "Ethics and Geoengineering: Reviewing the Moral Issues Raised by Solar Radiation Management and Carbon Dioxide Removal: Ethics & Geoengineering." *Wiley Interdisciplinary Reviews: Climate Change* 4, no. 1 (January 2013): 23–37. https://doi.org/10.1002/wcc.198.
Quinn, Daniel. *Ishmael*. New York: Bantam Books, 1995.
Quinn, Warren S. "Actions, Intentions, and Consequences: The Doctrine of Doing and Allowing." *The Philosophical Review* 98, no. 3 (July 1989): 287–312. https://doi.org/10.2307/2185021.
Rawls, John. *A Theory of Justice*. Cambridge, MA: Harvard University Press, 1999.
———. *Justice as Fairness: A Restatement*. Cambridge, MA: Harvard University Press, 2001.
———. *Political Liberalism*. New York: Columbia University Press, 2005.
———. *The Law of Peoples*. Cambridge, MA: Harvard University Press, 2002.

Rayner, Steve, Clare Heyward, Tim Kruger, Nick Pidgeon, Catherine Redgwell, and Julian Savulescu. "The Oxford Principles." *Climatic Change* 121, no. 3 (December 2013): 499–512. https://doi.org/10.1007/s10584-012-0675-2.

Reglitz, Merten. "Fairness to Non-Participants: A Case for a Practice-Independent Egalitarian Baseline." *Critical Review of International Social and Political Philosophy* 20, no. 4 (July 4, 2017): 466–85. https://doi.org/10.1080/13698230.2015.1037575.

Reynolds, Jesse. "A Critical Examination of the Climate Engineering Moral Hazard and Risk Compensation Concern." *The Anthropocene Review* 2, no. 2 (August 2015): 174–91. https://doi.org/10.1177/2053019614554304.

Robock, Alan. "20 Reasons Why Geoengineering May Be a Bad Idea." *Bulletin of the Atomic Scientists* 64, no. 2 (May 1, 2008): 14–18. https://doi.org/10.2968/064002006.

Rogelj, Joeri, Michel den Elzen, Niklas Höhne, Taryn Fransen, Hanna Fekete, Harald Winkler, Roberto Schaeffer, Fu Sha, Keywan Riahi, and Malte Meinshausen. "Paris Agreement Climate Proposals Need a Boost to Keep Warming Well below 2 °C." *Nature* 534, no. 7609 (June 29, 2016): 631–39. https://doi.org/10.1038/nature18307.

Rogelj, Joeri, Gunnar Luderer, Robert C. Pietzcker, Elmar Kriegler, Michiel Schaeffer, Volker Krey, and Keywan Riahi. "Energy System Transformations for Limiting End-of-Century Warming to below 1.5 °C." *Nature Climate Change* 5, no. 6 (May 21, 2015): 519–27. https://doi.org/10.1038/nclimate2572.

Royal Society. *Geoengineering the Climate: Science, Governance and Uncertainty.* London: The Royal Society, 2009.

Sangiovanni, Andrea. "Global Justice, Reciprocity, and the State." *Philosophy & Public Affairs* 35, no. 1 (2007): 3–39.

Scanlon, Thomas. *Being Realistic about Reasons.* Oxford: Oxford University Press, 2014.

———. "How I Am Not a Kantian." In *On What Matters*, by Derek Parfit. Oxford: Oxford University Press, 2011.

———. *What We Owe to Each Other.* Cambridge, MA: Belknap Press of Harvard University Press, 1998.

Scheffler, Samuel. "Doing and Allowing." *Ethics* 114, no. 2 (January 2004): 215–39. https://doi.org/10.1086/379355.

Schipper, E. Lisa F. "Conceptual History of Adaptation in the UNFCCC Process." *Review of European Community & International Environmental Law* 15, no. 1 (April 1, 2006): 82–92. https://doi.org/10.1111/j.1467-9388.2006.00501.x.

Schlosberg, David. *Defining Environmental Justice: Theories, Movements, and Nature.* Oxford: Oxford University Press, 2007.

Schmidt, Gavin. "30 Years after Hansen's Testimony." *Real Climate* (blog). Accessed July 1, 2018. http://www.realclimate.org/index.php/archives/2018/06/30-years-after-hansens-testimony/.

Schmidtz, David, and Elizabeth Willott, eds. *Environmental Ethics: What Really Matters, What Really Works.* 2nd edition. New York: Oxford University Press, 2012.

Shearer, Christine, Mick West, Ken Caldeira, and Steven J. Davis. "Quantifying Expert Consensus against the Existence of a Secret, Large-Scale Atmospheric Spraying Program." *Environmental Research Letters* 11, no. 8 (2016): 084011. https://doi.org/10.1088/1748-9326/11/8/084011.

Shepherd, John. "Napkin Diagram." Accessed August 18, 2017. http://jgshepherd.com/science-topics/geoengineering/.

Shrader-Frechette, Kristin. *Environmental Justice: Creating Equality, Reclaiming Democracy.* Oxford: Oxford University Press, 2002.

Shue, Henry. "Climate Surprises: Risk Transfers, Negative Emissions, and the Pivotal Generation." *SSRN Electronic Journal*, 2018. https://doi.org/10.2139/ssrn.3165064.

———. "Global Environment and International Inequality." *International Affairs* 75, no. 3 (July 1, 1999): 531–45. https://doi.org/10.1111/1468-2346.00092.

———. "Subsistence and Luxury Emissions." *Law and Policy* 15, no. 1 (1993): 39–59.

Simmons, A. J. *Justification and Legitimacy; Essays on Rights and Obligations.* Cambridge: Cambridge University Press, 2001.

Singer, Peter. "All Animals Are Equal." In *Environmental Ethics: What Really Matters, What Really Works*, edited by David Schmidtz and Elizabeth Willott, 2nd edition, 49–60. New York: Oxford University Press, 2012.
———. "One Atmosphere." In *Climate Ethics*, edited by Stephen Gardiner, Simon Caney, Dale Jamieson, and Henry Shue, 181–99. Oxford: Oxford University Press, 2010.
———. "Sidwick and Reflective Equilibrium." *The Monist* 58, no. 3 (1974): 490–517. https://doi.org/10.2307/27902380.
———. *The Expanding Circle: Ethics, Evolution, and Moral Progress*. 1st Princeton University Press pbk. ed. Princeton, NJ: Princeton University Press, 2011.
Sinnott-Armstrong, Walter. "It's Not My Fault: Global Warming and Individual Moral Obligations." In *Advances in the Economics of Environmental Resources*, edited by Richard Howarth, 5:285–307. Bingley, UK: Emerald Publishing, 2005. https://doi.org/10.1016/S1569-3740(05)05013-3.
SRMGI. "Solar Radiation Management Governance Initiative." Accessed August 18, 2017. http://www.srmgi.org/.
Steinbock, Bonnie. "Moral Status, Moral Value, and Human Embryos: Implications for Stem Cell Research." In *The Oxford Handbook of Bioethics*, edited by Bonnie Steinbock. Oxford: Oxford University Press, 2007.
———, ed. *The Oxford Handbook of Bioethics*. Oxford Handbooks. Oxford: Oxford University Press, 2007.
Stern, Nicholas. *The Economics of Climate Change: The Stern Review*. Edited by Great Britain. Cambridge: Cambridge University Press, 2007.
Stevens, Greg A., and James Burley. "3,000 Raw Ideas = 1 Commercial Success!" *Research-Technology Management* 40, no. 3 (1997): 16–27.
Stocker, Michael. "The Schizophrenia of Modern Ethical Theories." *Journal of Philosophy* 73, no. 14 (1976): 453–66. https://doi.org/10.2307/2025782.
Stone, Christopher. "Should Trees Have Standing?" In *Environmental Ethics: What Really Matters, What Really Works*, edited by David Schmidtz and Elizabeth Willott, 2nd edition, 85–89. New York: Oxford University Press, 2012.
Sunstein, Cass R. *Laws of Fear: Beyond the Precautionary Principle*. Cambridge: Cambridge University Press, 2005.
Svoboda, Toby. *The Ethics of Climate Engineering: Solar Radiation Management and Non-Ideal Justice*. New York: Routledge, 2017.
———. "The Ethics of Geoengineering: Moral Considerability and the Convergence Hypothesis: The Ethics of Geoengineering." *Journal of Applied Philosophy* 29, no. 3 (August 2012): 243–56. https://doi.org/10.1111/j.1468-5930.2012.00568.x.
Svoboda, Toby, Peter J. Irvine, Daniel Edward Callies, and Masa Sugiyama. "The Potential for Climate Engineering with Stratospheric Sulfate Aerosol Injections to Reduce Climate Injustice." *Journal of Global Ethics*, forthcoming (n.d.).
Szerszynski, Bronislaw, Matthew Kearnes, Phil Macnaghten, Richard Owen, and Jack Stilgoe. "Why Solar Radiation Management Geoengineering and Democracy Won't Mix." *Environment and Planning A* 45, no. 12 (December 2013): 2809–16. https://doi.org/10.1068/a45649.
Taylor, Dorceta. *Toxic Communities: Environmental Racism, Industrial Pollution, and Residential Mobility*. New York: New York University Press, 2014.
Taylor, Paul W. *Respect for Nature: A Theory of Environmental Ethics*. 25th anniversary edition. Princeton, NJ: Princeton University Press, 2011.
"The European Transdisciplinary Assessment of Climate Engineering (EuTRACE)." Accessed August 17, 2017. https://www.adelphi.de/en/publication/european-transdisciplinary-assessment-climate-engineering-eutrace.
Thomas, Chris D., Alison Cameron, Rhys E. Green, Michel Bakkenes, Linda J. Beaumont, Yvonne C. Collingham, Barend F. N. Erasmus, et al. "Extinction Risk from Climate Change." *Nature* 427, no. 6970 (January 8, 2004): 145–48. https://doi.org/10.1038/nature02121.
Thomson, Judith Jarvis. "A Defense of Abortion." *Philosophy & Public Affairs* 1, no. 1 (1971): 47–66.

———. "Killing, Letting Die, and the Trolley Problem." Edited by Sherwood J. B. Sugden. *Monist* 59, no. 2 (1976): 204–17. https://doi.org/10.5840/monist197659224.

———. "The Trolley Problem." *The Yale Law Journal* 94, no. 6 (May 1985): 1395. https://doi.org/10.2307/796133.

———. "Turning the Trolley." *Philosophy & Public Affairs* 36, no. 4 (2008): 359–74.

Tremmel, Jörg, ed. *Handbook of Intergenerational Justice*. Cheltenham, UK: Edward Elgar, 2006.

Trisos, Christopher H., Giuseppe Amatulli, Jessica Gurevitch, Alan Robock, Lili Xia, and Brian Zambri. "Potentially Dangerous Consequences for Biodiversity of Solar Geoengineering Implementation and Termination." *Nature Ecology & Evolution* 2, no. 3 (March 2018): 475–82. https://doi.org/10.1038/s41559-017-0431-0.

Turner, Jack. *The Abstract Wild*. 3. [ed.]. Tucson: University of Arizona Press, 1999.

Tyler, Tom R. *Why People Obey the Law*. New Haven, CT: Yale University Press, 1992.

U. S. Government Accountability Office. "Climate Change: Preliminary Observations on Geoengineering Science, Federal Efforts, and Governance Issues," no. GAO-10-546T (March 18, 2010). http://www.gao.gov/products/GAO-10-546T.

United Nations. "Convention On Biological Diversity," 1992. http://unfccc.int/essential_background/convention/items/6036.php.

———. "Paris Agreement," 2015. https://unfccc.int/process-and-meetings/the-paris-agreement/the-paris-agreement.

———. "Rio Declaration on the Environment and Development," August 1992. http://www.un.org/documents/ga/conf151/aconf15126-1annex1.htm.

———. "United Nations Framework Convention on Climate Change," 1992. http://unfccc.int/essential_background/convention/items/6036.php.

———. "World Charter for Nature," 1982. http://www.un.org/documents/ga/res/37/a37r007.htm.

United Nations Development Program, ed. *Work for Human Development*. Human Development Report 2015. New York: United Nations Development Program, 2015. hdr.undp.org/sites/default/files/2015_human_development_report.pdf.

Valentini, Laura. "Ideal vs. Non-Ideal Theory: A Conceptual Map." *Philosophy Compass* 7, no. 9 (September 1, 2012): 654–64. https://doi.org/10.1111/j.1747-9991.2012.00500.x.

Victor, D. G. "On the Regulation of Geoengineering." *Oxford Review of Economic Policy* 24, no. 2 (June 1, 2008): 322–36. https://doi.org/10.1093/oxrep/grn018.

Victor, David G. *Global Warming Gridlock: Creating More Effective Strategies for Protecting the Planet*. Cambridge: Cambridge University Press, 2011.

Virgoe, John. "International Governance of a Possible Geoengineering Intervention to Combat Climate Change." *Climatic Change* 95, no. 1–2 (July 1, 2009): 103–19. https://doi.org/10.1007/s10584-008-9523-9.

Walzer, Michael. *Just and Unjust Wars: A Moral Argument with Historical Illustrations*. Fifth edition. New York: Basic Books, 2015.

Weber, Max. *The Theory of Social and Economic Organization*. Translated by Talcott Parsons. New York: Free Press, 1997.

Weinberg, Alvin. *Reflections on Big Science*. Cambridge, MA: The MIT Press, 1967.

Weisbach, David. "Negligence, Strict Liability, and Responsibility for Climate Change | Belfer Center for Science and International Affairs." *The Harvard Project on International Climate Agreements*, 2010.

Whitman, Jeffrey P. "The Many Guises of the Slippery Slope Argument:" *Social Theory and Practice* 20, no. 1 (1994): 85–97. https://doi.org/10.5840/soctheorpract19942012.

Williams, Bernard. "Which Slopes Are Slippery?" In *Moral Dilemmas in Modern Medicine*, edited by Michael Lockwood. Studies in Bioethics. London: Oxford University Press, 1985.

Wong, Pak-Hang. "Consenting to Geoengineering." *Philosophy & Technology* 29, no. 2 (June 2016): 173–88. https://doi.org/10.1007/s13347-015-0203-1.

Woods, Ngaire. "Holding Intergovernmental Institutions to Account." *Ethics and International Affairs* 17, no. 1 (2003): 69–80.

World Bank. "CO2 Emissions (Metric Tons per Capita) | Data." Accessed August 18, 2017. https://data.worldbank.org/indicator/.

World Health Organization. "Ambient Outdoor Air Quality and Health." Accessed August 18, 2017. http://www.who.int/mediacentre/factsheets/fs313/en/.
World Resources Institute. "CAIT Climate Data Explorer." Accessed August 18, 2017. http://cait.wri.org/.
———. "Per Capita CO2 Emissions for Select Major Emitters, 2007 and 2030 (Projected) | World Resources Institute." Accessed August 18, 2017. http://www.wri.org/resources/charts-graphs/capita-co2-emissions-select-major-emitters-2007-and-2030-projected.
Wyk, Jo-Ansie van. "Atoms, Apartheid, and the Agency: South Africa's Relations with the IAEA, 1957–1995." *Cold War History* 15, no. 3 (July 3, 2015): 395–416. https://doi.org/10.1080/14682745.2014.897697.

Index

ability to pay principle, 93, 106–109
accountability, 39, 82–83, 123
adaptation, 1, 17, 28, 41, 42, 44–45, 50n91, 69, 108, 113n57, 139
afforestation, 5
albedo, 5, 17n2, 95
All Affected Principle, 116, 123–125
anthropocentrism, 142–145
anthropogenic climate change, 1, 6, 29, 31, 40, 41–42, 54, 61, 71, 96, 98–99, 103–104, 144, 147
anti-lock brakes, 43–44
Aquinas, St. Thomas, 53
arm the future argument, 23–28
Arneson, Richard, 120
atmospheric concentration of GHGs, 2, 25–26, 29, 30, 72n22, 98–99, 104, 112n52, 137–138
authoritarianism, 28, 39
autonomy, 111n15, 124, 135n18

BECCS. *See* bioenergy with carbon capture and storage
Beitz, Charles, 91n43, 125–126, 135n24
beneficiary pays principle, 103, 112n50, 113n53
beneficiary responsibility. *See* responsibility
Bentham, Jeremy, 73n43, 143, 150n21
bioenergy with carbon capture and storage (BECCS), 138

biocentrism, 58–59, 143–144
biodiversity, 2, 5, 61, 96, 144, 150n26
buying time, 38, 45

Callicott, J. Baird, 145
carbon dioxide removal (CDR), 4–5, 29, 138, 139
carbon emissions, 30, 45, 67, 98–99, 102, 103–104, 112n52, 138, 149n8
catastrophe, 13, 24, 25, 28, 55, 91n37
causal responsibility. *See* responsibility
CBD. *See* Convention on Biological Diversity
CDR. *See* carbon dioxide removal
CFCs. *See* chlorofluorocarbons
chemtrails, 4
China, 94–95, 99, 131, 136n36
chlorofluorocarbons (CFCs), 24, 30
climate change, 1; anthropogenic, 1, 4, 19n28, 29, 98–93; denial of, 26; natural, 61, 98
climate engineering, 1; cost, 7, 19n28, 20n46, 28–31, 48n28; definition, 1; delivery methods, 6; different kinds of, 4–7; field testing of, 33, 35, 45, 46, 84; synonyms, 7; side effects, 5, 7, 12, 24, 29, 30–31, 38, 57, 62, 71, 73n41, 95, 144
cloning, 37
coercion, 76, 88, 89n16, 96, 111n15
cognitive diversity, 121

comparative benefit, 81–82, 89n16, 90n20, 90n23, 91n37
consent, 19n41, 76–77, 88n4, 91n36
Convention on Biological Diversity (CBD), 53, 72n5
convergence hypothesis, 20n58, 144
Crutzen, Paul, 11, 17n2, 23, 25, 26, 47n8

Daniels, Norman, 9–10, 19n31, 19n34
democracy, 13, 18n4, 39, 49n69, 78, 86, 118, 120–123, 123–127, 130, 135n19, 135n22
developed countries, 73n40, 98, 100, 102–104, 106–108
developing countries, 44, 73n40, 98–99, 102, 104–105, 106–108, 109, 111n31, 112n42, 127
difference principle, 119, 129
dilemma. *See* moral dilemma
doctrine of doing and allowing, 63–66
doctrine of double effect, 66–70
doing and allowing. *See* doctrine of doing and allowing
domination, 59–61
double effect. *See* doctrine of double effect

ecocentrism, 143–146
economic argument, 7, 28–31
equal influence principle, 125–126
equity, 93, 96–97, 106
Europe, 99, 111n35
expanding circle, 143–146

fairness. *See* equity
fair terms: of inclusion, 115, 127; of participation, 115, 127
fault, 100–101; no-fault, 100–101
feasibility, 127; political, 128–129; theoretical, 128–129
future generations, 3, 13, 24–25, 26, 27, 76, 97, 130, 139–142

Gardiner, Steven, 12, 24–28, 32, 56–57, 72n24, 98, 130, 139
GDP. *See* gross domestic product
geoengineering. *See* climate engineering; stratospheric aerosol injection; carbon dioxide removal
GHG. *See* greenhouse gases

global mean surface temperature, 2, 7, 25, 29–30, 57, 73n38, 98, 144
Gore, Albert, 27
governance, 2, 12, 27, 75–76, 78, 80–87
greenhouse gases (GHG), 54, 105, 138; atmospheric concentration of, 26, 138
Greenland Ice Sheet, 61, 73n38
gross domestic product (GDP), 19n28, 104, 107, 108

Hale, Benjamin, 42, 43, 49n81, 150n23
Hansen, James, 27, 137
Hardin, Garret, 3
HDI. *See* Human Development Index
Heyward, Clare, 13, 19n30, 20n59, 113n53, 135n22
historical responsibility. *See* responsibility
hubris, 17, 34
Human Development Index (HDI), 104, 107–108, 109, 112n48, 113n55

ideal theory, 10–11, 19n37, 19n39, 128–129
inclusion. *See* fair terms of inclusion
injustice, 71, 86, 96, 105, 109, 112n52, 142
instrumental value of fair procedures, 115–122
international cooperation, 3, 88n2, 96
intergenerational justice. *See* justice
Intergovernmental Panel on Climate Change (IPCC), 2, 5, 25, 26, 47n10, 47n14, 61, 102, 138
intrinsic value of fair procedures, 122–123
intuitionism, 8
IPCC. *See* Intergovernmental Panel on Climate Change

Jamieson, Dale, 12, 20n52, 32, 34, 37, 52, 58, 59–60, 72n27, 146
justice, 84, 89n5; concept of, 84, 89n5, 96; intergenerational, 24, 139–142; and legitimacy, 85, 90n21; natural duty of, 122–123; procedural. *See* procedural justice; substantive, 84–85, 93–94, 96–98, 109
just savings principle, 141

Keith, David, 18n24, 29, 31, 38, 42, 46, 47n22, 71, 91n39

Kyoto Protocol, 3

legitimacy, 10, 19n37, 27, 75–76; concept of, 76, 77–78; descriptive versus normative, 8, 84, 87, 89n14, 91n45; Metacoordination View of, 78, 87, 89n14, 91n38; normative criteria of, 78–87; Rawlsian conception of, 76–77; voluntarist conception of, 76–77
Leopold, Aldo, 144, 145
lock-in. *See* slippery slope argument

maximin, 56, 57, 72n18, 72n24
Metacoordination View. *See* legitimacy
methodology. *See* reflective equilibrium; ideal theory; nonideal theory
Mill, John Stuart, 136n40, 140
minimax, 56, 57, 72n18, 72n24
mitigation, 1, 4, 7, 17, 19n28, 23, 25, 27, 28, 28–31, 42, 56, 69, 93, 113n55, 136n39, 139
moderate proceduralism, 118
Moellendorf, Darrel, 47n8, 56, 57, 59, 72n24, 96–98, 100, 111n17
monsoon, 29, 39–40
Montreal Protocol, 30
moral : considerability, 20n58, 143–144; corruption, 24–25, 26, 27, 111n26; dilemma, 13, 20n57, 25, 70–71; hazard. *See* moral hazard argument; reasons, 34, 38–39; responsibility; responsibility; schizophrenia, 26, 47n16; theory; reflective equilibrium
moral hazard argument, 14, 41–45
Morrow, David, 12, 20n56, 64, 65–66, 73n44, 88n4, 91n35
Mount Pinatubo, 6–7

National Oceanic and Atmospheric Administration (NOAA), 48n36, 138, 149n2
natural climate change. *See* climate change
natural duty of justice. *See* justice
Natural Environment Research Council, 42–43, 49n86
ND-GAIN. *See* Notre Dame Global Adaptation Index
negative emissions technology (NET). *See* carbon dioxide removal

nightmare scenario, 25–26, 28
NOAA. *See* National Oceanic and Atmospheric Administration
no-fault. *See* fault
nonideal theory, 10–11, 19n40, 137–139
Nordhaus, William, 19n28, 29, 48n25
normative criteria. *See* legitimacy
normative gap, 59
normative theorizing. *See* reflective equilibrium; ideal theory; nonideal theory
Notre Dame Global Adaptation Index (ND-GAIN), 108, 113n57, 131–132
Nozick, Robert, 118, 134n6, 136n30
nuclear energy, 36–37, 49n59
Nussbaum, Martha, 70, 74n67

ocean acidification, 7, 30, 112n52
ocean fertilization, 5
Oxford Principles, 12, 20n51
ozone depletion, 30, 48n36, 110n7; effects of SAI on, 7, 30–31, 40, 95–96

Parfit, Derek, 38, 49n67
Paris Agreement, 3, 18n12, 18n13, 18n15; insufficiency of, 4, 18n14, 25
participation. *See* fair terms of participation
path-dependency. *See* slippery slope argument
per capita emissions, 99, 102, 104, 111n36, 112n38, 112n42
Pinatubo. *See* Mount Pinatubo
playing God argument, 15, 61–66, 70, 71
political feasibility. *See* feasibility
political legitimacy. *See* legitimacy
polluter pays principle. *See* causal responsibility
precautionary argument, 51, 52–58
precautionary principle, 52–56, 72n1, 72n7, 72n9; Rawlsian Core Precautionary Principle, 56–57, 72n20
precipitation: effects of SAI on, 29, 40, 94–95, 144
Preston, Christopher, 13, 21n61
principle of postponement, 139, 149n11
procedural justice, 85–87, 91n44, 115–116; imperfect, 117; perfect, 117; pure, 118; quasi-pure, 119

Proportionality Principle, 116, 126–127, 127–129, 131–133
public engagement, 8, 18n16, 46, 49n86, 86, 91n36
pure instrumentalism, 120

Rawls, John, 8, 10, 19n32, 19n38, 19n39, 56, 72n24, 76–77, 115, 117–118, 119, 122, 139, 139–142
reciprocity, 96–97, 111n18
reflective equilibrium, 8–10, 19n31; narrow, 8–9; wide, 9–10, 19n34, 19n36, 65
respect, 58, 78, 89n13, 110n14, 111n17, 122, 127
respect for nature argument, 15, 52, 58–61
responsibility: beneficiary, 103–105, 112n50, 113n53; causal, 16, 98–102; moral, 98, 99–101; outcome, 111n32
Rio Declaration, 53, 72n4
risk compensation, 42, 49n78
Royal Society, 4, 12, 17n1

SAI. *See* stratospheric aerosol injection
sentience, 143
Shue, Henry, 93, 97, 98, 103, 110n2, 112n37, 139, 149n10
simple view, 125–126
Singer, Peter, 19n36, 104, 112n44, 143–144, 149n18, 150n20
slippery slope argument, 14, 31–41; modest, 34; decisive, 35
solar radiation management (SRM), 5, 5–7
Solar Radiation Management Governance Initiative (SRMGI), 50n95
space mirrors, 6

SRM. *See* solar radiation management
SRMGI. *See* Solar Radiation Management Governance Initiative
stage gates, 45, 47, 50n96
stratospheric aerosol injection (SAI), 6–8; cost, 6–7, 28–31; delivery methods, 6; natural analog of, 6–7; side effects, 30–31, 38
substantive justice. *See* justice
Svoboda, Toby, 13, 20n58, 21n63, 110n1, 111n20, 144

Taylor, Paul, 58–59, 73n28, 143–144
technological fix, 25
termination problem, 56, 144
theoretical feasibility. *See* feasibility
tragedy of the commons, 2–3
transparency, 39, 79, 83–84, 91n32, 91n33, 91n35, 91n36

UNFCCC. *See* United Nations Framework Convention on Climate Change
unilateral deployment, 75, 81, 88n2
United Nations Framework Convention on Climate Change (UNFCCC), 47n17, 52–53, 77, 97, 98, 138
United States, 4, 95, 99, 108, 137
unknown unknowns, 7, 31
utilitarianism, 19n36, 63, 73n43, 140–141

vulnerability, 44, 108–109

wealth, 73n40, 98–101, 101–104, 105, 106–108
World Health Organization, 73n37

About the Author

Daniel Edward Callies is a postdoctoral scholar at the University of California, San Diego's Institute for Practical Ethics. His research generally focuses on issues at the intersection of ethics, politics, the environment, and technology. He received his BA and MA in philosophy from San Diego State University, and his PhD from Johann Wolfgang Goethe Universität Frankfurt (Germany). Prior to joining UCSD, he was a predoctoral research fellow at Harvard University's Kennedy School of Government and the Bernheim Postdoctoral Fellow in Social Responsibility at Université catholique de Louvain (Belgium). He has articles appearing in the *Journal of Applied Philosophy*; *Global Environmental Politics*; *Moral Philosophy and Politics*; the *Journal of Global Ethics*; and *Ethics, Policy & Environment*. Additionally, Daniel heads the energy and environment research cluster for the Agenda for International Development. On weekends, he can be found surfing the many beaches on offer in Southern California.

www.ingramcontent.com/pod-product-compliance
Lightning Source LLC
Chambersburg PA
CBHW050908300426
44111CB00010B/1426